OT 30
Operator Theory: Advances and Applications
Vol. 30

Birkhäuser Verlag
Basel · Boston · Berlin

Yurii I. Lyubich

Introduction to the Theory of Banach Representations of Groups

Translated from the Russian
by A. Jacob

1988

Birkhäuser Verlag
Basel · Boston · Berlin

Author's address:
Prof. Yurii I. Lyubich
Khar'kovskii Universitet
pl. Dzerzinskogo 4
Khar'kov 310077
USSR

Translation of:
Vvedenie v teoriyu banakhovykh
predstavlenii grupp, »Vyshcha Shkola«,
Khar'kov, 1985

Library of Congress Cataloging in Publication Data

Liubich, I͡U. I. (I͡Urii Il′ich)
   [Vvedenie v teorii͡u banakhovykh predstavlenii͡ grupp. English]
   Introduction to the theory of Banach representations of groups /
Yurii I. Lyubich ; translated from the Russian by A. Jacob.
   p. cm. – – (Operator theory, advances and applications ; vol. 30)
   Translation of: Vvedenie v teorii͡u banakhovykh predstavlenii͡ grupp.
   Bibliography: p.
   Includes index.
   ISBN 0–8176–2207–1 (U.S.)
   1. Locally compact groups.   2. Representations of groups.
3. Banach algebras.   I. Title.   II. Series: Operator theory.
advances and applications ; v. 30.
QA387.L5813   1988
512′.55 – – dc19

CIP-Kurztitelaufnahme der Deutschen Bibliothek

Ljubič, Jurij I.:
Introduction to the theory of Banach representations of groups
/ Yurii I. Lyubich. Transl. from the Russ. by A. Iacob. – Basel ;
Boston ; Berlin : Birkhäuser, 1988
   (Operator theory ; Vol. 30)
   Einheitssacht.: Vvedenie v teoriju banachovych predstavlenij grupp
   <engl.>
   ISBN 3–7643–2207–1 (Basel ...) Pb.
   ISBN 0–8176–2207–1 (Boston) Pb.
NE: GT

© 1988 Birkhäuser Basel
Printed in Germany
ISBN 3-7643-2207-1
ISBN 0-8176-2207-1

CONTENTS

PREFACE

The theory of group representations plays an important role in modern mathematics and its applications to natural sciences. In the compulsory university curriculum it is included as a branch of algebra, dealing with representations of finite groups (see, for example, the textbook of A. I. Kostrikin [25]). The representation theory for compact, locally compact Abelian, and Lie groups is covered in graduate courses, concentrated around functional analysis. The author of the present book has lectured for many years on functional analysis at Khar'kov University. He subsequently continued these lectures in the form of a graduate course on the theory of group representations, in which special attention was devoted to a retrospective exposition of operator theory and harmonic analysis of functions from the standpoint of representation theory. In this approach it was natural to consider not only unitary, but also Banach representations, and not only representations of groups, but also of semigroups.

The first work on the algebraic theory of semigroups was written in 1928 by the Khar'kov mathematician A. K. Sushkevich (Suschkewitsch). The object that he discovered (the kernel or "kerngruppe" of the semigroup, presently known as the Sushkevich kernel) was later revealed not only in finite semigroups, but also, for example, in compact semigroups. The Sushkevich kernel of a compact semigroup is in a number of important instances a compact group, and in the general case it decomposes into mutually isomorphic compact groups. This reduces, to a considerable extent, the study of compact semigroups to the group situation. Following this path and by using also the notion of Bohr compactification, K. de Leeuw and I. Glicksberg obtained a general structure theorem for (weakly) almost periodic operator semigroups in 1961. Their result soon found important applications in probability theory. The outcome is that today one has a theory of almost periodic representations of semigroups with applications to the generalized Perron-Frobenius theory (i.e., the theory of nonnegative operators and representations); from there applications were found to the theory of dynamical systems, Markov chains, and so on. This circle of problems is considered in Chapter 4, devoted to the classical theory of representations of compact groups (the Peter-Weyl theory) and its applications. This chapter contains also a detailed study of almost periodic functions on groups and semigroups. From a spectral viewpoint, this is the theory of the discrete spectrum.

The continuous spectrum arises naturally in the context of locally compact Abelian groups. Banach (or nonunitary Hilbert) representations decompose according to characters, which form a continuous spectrum only in the "Pickwickian" sense. In the attempt to make the latter precise one is led to the notion of separable spectrum around which Chapter 5 concentrates.

Each of Chapters 4 and 5 may serve as a short graduate course. The preceding chapters contain background material on the spectral theory of operators and Banach algebras (Chapter 1), elements of the theory of topological groups and semigroups (Chapter 2), and the beginning of representation theory (Chapter 3). A considerable part of Chapter 1 should be covered by a general course in functional analysis. However, on the whole we assume that the reader is prepared in this domain, as well as in other branches of mathematics which are part of the university curriculum. The prerequisites (terminology and facts) for the book can be found in the textbooks indicated in the list of references. This list has basically a "historical" character, and may also serve in enlarging the reader's horizon.

Modern representation theory is exceptionally vast and rich. And despite the fact that the available textbooks and monographs (among which me mention here those of A. A. Kirillov [23], D. P. Zhelobenko [50], and E. Hewitt and K. A. Ross [19]) cover a very large part of this theory, they do not treat sufficiently Banach representations and the case of semigroups, i.e., the main aspects considered here. Unfortunately, lack of space has forced the author to avoid considering unbounded operators. As a consequence, the theory of infinitesimal operators remains beyond the limits of this exposition, though we treat here strongly continuous, and not only uniformly (norm) continuous representations. Regarding the subject of infinitesimal operators we refer the reader to the monograph of E. Hille and R. Phillips [20].

We included a sufficient number of (mostly easy) exercises. As a rule, they must be solved if one wishes to master the material treated in the book.

The author is deeply grateful to M. G. Krein and Yu. A. Drozd, who, in the quality of referees of the manuscript, made a number of useful remarks.

CHAPTER 1

ELEMENTS OF SPECTRAL THEORY

1. INTEGRATION OF VECTOR-VALUED FUNCTIONS

1°. We give an exposition of the analytic apparatus that will be used systematically in this text. It was developed in the thirties by S. Bochner, I.M. Gelfand, and other mathematicians.

Let S be an arbitrary nonempty set and let B be a Banach space (unless otherwise stipulated, all Banach spaces considered here are complex, i.e., the ground field is $\mathbb{C}$, the complex numbers; also, the letter B always designates a Banach space, and we use H to denote Hilbert spaces). A map $X:S \to B$ is called a *vector-valued function* or a *vector-function* (on S with values in B). With each vector-function X one associates the nonnegative scalar function $s \to \|X(s)\|$ $(s \in S)$. If it is bounded, then the vector-function X is also said to be *bounded*. For a bounded vector-function X one defines the norm $\|X\|$ by $\|X\| = \sup_s \|X(s)\|$. The set $B(S,B)$ of bounded B-valued vector-functions on S is a Banach space with respect to the pointwise operations of addition and multiplication by a scalar, and to the norm introduced above. The convergence of a sequence

$\{X_k\} \subset B(S,B)$ means its uniform convergence. If S is a topological vector space one can speak about continuous B-valued vector-functions on S. The bounded continuous functions form a subspace $CB(S,B)$ of $B(S,B)$. If S is compact every continuous vector-function on S is bounded (and attains its norm). In this case $CB(S,B)$ coincides with the Banach space $C(S,B)$ of all continuous vector-functions on S. In the case $B = \mathbb{C}$ we deal with scalar functions and the corresponding vector spaces are denoted simply by $B(S)$, $CB(S)$, and $C(S)$. Suppose S is a locally compact topological space on which there is given a measure ds (all measures considered in this monograph are assumed to be Borelian and regular). Then with respect to this measure one can integrate also vector functions, and not only scalar ones. This circumstance plays a very important role in applications to representation theory, to which this book is devoted. However, for our purposes it is not necessary to develop here the general theory of integration of vector-functions (the reader interested in the latter is referred to the relevant volume of Bourbaki's treatise [4]). It will suffice to consider integrals of the form $\int_S \phi(s)X(s)ds$, where $\phi \in L_1(S;ds)$ (i.e., ϕ is a ds-summable scalar function on S) and $X \in CB(S,B)$. As a rule we shall simply write $\int$ instead of $\int_S$ (but not for integrals over subsets $M \subset S$).

THEOREM-DEFINITION. *For every scalar function* $\phi \in L_1(S;ds)$ *and every bounded weakly-continuous vector-function* X *on* S *there exists a unique element* $x \in B$ *such that*

$$f(x) = \int \phi(s)f(X(s))ds \tag{1}$$

for all $f \in B^*$, *i.e., for all linear functionals on* B. x *is called the integral of the product* $\phi(s)X(s)$ *with respect to the measure* ds *and is accordingly denoted by* $x = \int \phi(s)X(s)ds$.

[By "linear functional" we shall always mean "continuous additive and homogeneous functional".]

PROOF. If the required element x exists, then it is unique, since equality (1) uniquely specifies the value $f(x)$ for all $f \in B^*$. Turning to the proof of existence, we remark that the identity $\xi(f) = \int \phi(s) f(X(s)) ds$ defines a linear functional ξ on B^* (indeed, $\|\xi\| \leq \int |\phi(s)| \, \|X(s)\| \, ds < \infty$), i.e., an element of the Banach space $B^{**} \supset B$. This completes the proof in the case where B is reflexive. Even if B is not reflexive, it is nevertheless closed in B^{**}, and to prove the theorem it suffices to approximate ξ arbitrarily well by elements of B.

As S is locally compact, for every $\varepsilon > 0$ there is a compact set $K \subset S$ such that $\int_{S \setminus K} |\phi(s)| \, \|X(s)\| \, ds < \varepsilon$. Hence, it suffices to prove that $\xi \in B$ when S is compact. With no loss of generality we can restrict ourselves to the following setting: ds is a probability measure, $\phi = 1\!\!1$ (the function identically equal to 1), and the space B is real.

Pick an arbitrary finite set $F = \{f_1, \ldots, f_n\} \subset B^*$ and consider the map $\phi\colon S \to \mathbb{R}^n$ defined by the formula $\phi(s) = (f_1(X(s)), \ldots, f_n(X(s)))$. Its image $\operatorname{Im} \Phi$ is compact, and hence so is its convex hull $C_F = \operatorname{Co}(\operatorname{Im} \Phi)$. We show that the point $\xi_F = (\xi(f_1), \ldots, \xi(f_n))$ belongs to C_F. In fact, otherwise there exists a half-space $\Sigma_{k=1}^n \alpha_k \xi_k > \beta$ which contains C_F but not ξ_F. Then $\Sigma_{k=1}^n \alpha_k \xi(f_k) < \Sigma_{k=1}^n \alpha_k f_k(X(s))$ for all s. Integrating this inequality we arrive at a contradiction, because $\xi(f_k) = \int_S f_k(X(s)) ds$, $1 \leq k \leq n$. Thus, $\xi_F \in C_F$, i.e., there exist a finite set $\{s_1, \ldots, s_m\} \subset S$ and numbers $p_1, \ldots, p_m$ $(p_i \geq 0, \ \Sigma_i \, p_i = 1)$, such that $\xi(f_k) = \Sigma_{i=1}^m p_i f_k(X(s_i))$. Setting $x_F = \Sigma_{i=1}^m p_i X(s_i) \in B$, we get $\xi(f_k) = f_k(x_F)$, $1 \leq k \leq n$. Now notice that $x_F \in \operatorname{Co}(X(S))$ and the range $X(C) \subset B$ is weakly compact. Consequently, by a well known theorem of M.G. Krein and B.L. Shmul'yan (the proof of which can be found, for example, in M. Day's book [11]), the weak closure Q of the convex hull of X(S) is weakly compact. Since $x_F \in Q$, it follows that the set

$M_F = \{x \mid x \in Q,\ f_k(x) = \xi(f_k),\ 1 \leqslant k \leqslant n\}$ is nonempty and, obviously, weakly compact. Next, since $M_{F_1} \cap \ldots \cap M_{F_r} = M_{F_1 \cap \ldots \cap F_r} \neq \emptyset$, the family $\{M_F\}$ of compact sets is centered, and so $\bigcap_F M_F \neq \emptyset$. But then the vector $x \in \bigcap_F M_F$ satisfies the condition $\xi(f) = f(x)$ for all $f \in B^*$. Consequently, $\xi = x \in B$.

□

Remark. The bound

$$\left\| \int \phi(s)X(s)ds \right\| \leqslant \int |\phi(s)|\ \|X(s)\| ds$$

shows that *the mapping* $\phi \to \int \phi(s)X(s)ds$ *is a Banach space morphism* $L_1(S;ds) \to B$ (*its norm is* $\leqslant \|X\|$), *and the mapping* $X \to \int \phi(x)X(s)ds$ *is a morphism* $CB(S) \to B$ (*its norm is* $\leqslant |\phi|$).

[Recall that a *Banach space morphism* $B_1 \to B_2$ is a continuous linear mapping of B_1 into B_2. This terminology suits the modern treatment of the class of Banach spaces as a category.]

2°. An important particular case is the integration of holomorphic vector-functions along paths in the complex plane. A B-valued vector-function $X(\lambda)$ defined in a domain $D \subset \mathbb{C}$ is said to be *holomorphic* in D if every point $\lambda_o \in D$ has a neighborhood $|\lambda-\lambda_o| < \delta$, contained in D, in which $X(\lambda)$ admits a (convergent) series expansion $\Sigma_{k=0}^{\infty} a_k(\lambda-\lambda_o)^k$ (with coefficients $a_k \in B$). The next basic lemma completely reduces the complex analysis of vector-functions to the standard theory of scalar functions of one complex variable).

LEMMA. *In order for the continuous vector-function* $X: D \to B$ *to be holomorphic in* D *it is necessary and sufficient that it be weakly (or scalarly) holomorphic in* D, *i.e., that every scalar function* $f(X(\lambda))$ *with* $f \in B^*$ *be holomorphic in* D.

PROOF. The necessity of this condition is obvious. To prove its sufficiency, pick a point $\lambda_o \in D$, a disk $|\lambda-\lambda_o| < \delta$ contained in D, and an $f \in B^*$, and consider the Taylor expansion $f(X(\lambda)) = \Sigma_{k=0}^{\infty} \alpha_k(f)(\lambda-\lambda_o)^k$. Its coefficients can be calculated by Cauchy's integral formula:

$$\alpha_k(f) = \frac{1}{2\pi i}\int_{|\zeta-\lambda_o|=\rho} (\zeta-\lambda_o)^{-k-1} f(X(\zeta))d\zeta,$$

with arbitrary $0 < \rho < \delta$. By the theorem on the existence of the integral of a vector-function, $\alpha_k(f) = f(a_k)$, where a_k are vectors in B. Moreover, $\|a_k\| \leq M(\rho)\rho^{-k}$, where $M(\rho) = \max_{|\zeta-\lambda_o|=\rho} \|X(\zeta)\|$. It follows from these estimates that the power series $\Sigma_{k=0}^{\infty} a_k(\lambda-\lambda_o)^k$ converges in the disk $|\lambda-\lambda_o| < \rho$ for every $\rho < \delta$, and consequently in the disk $|\lambda-\lambda_o| < \delta$. Its sum is equal to $X(\lambda)$, because

$$f(X(\lambda)) = \Sigma_{k=0}^{\infty} f(a_k)(\lambda-\lambda_o)^k$$

for all $f \in B^*$.

□

We propose the reader to follow independently the details of the extension of complex analysis to the vector case. The main landmarks on this path are Cauchy's theorem on the integral along a closed contour, Cauchy's integral formula, the theory of isolated singularities (based on the consideration of Laurent's series), and Liouville's theorem asserting that any bounded entire function is constant. To these one should add differential calculus in the complex plane (which in standard treatments is customarily the first and foremost topic). In connection with this we mention that the definition of the derivative of a vector-function of one complex (or real) variable does not differ in form from the classical definition, but, needless to say, the corresponding passage to the limit is carried out in the Banach space B. Using the basic lemma given above it is readily verified that a vector-function is holomorphic in a domain D if and only if it is differentiable throughout D, which in turn implies (thanks to Cauchy's formula) that it is infinitely differentiable and that its derivatives admit the respective integral representation. The coefficients of the power series for $X(\lambda)$ in the neighborhood of the point λ_0 are given by the classical formulas $a_k = (k!)^{-1}X^{(k)}(\lambda_0)$, i.e., the series is identical to the Taylor series of $X(\lambda)$. We mention also that its radius of convergence is equal to

$$\rho = (\overline{\lim_{k\to\infty}} \|a_k\|^{1/k})^{-1} .$$

In the scalar case this is the well-known *Cauchy-Hadamard formula*, and since in the vector case it is proved in exactly the same way (only replacing the modulus by norm), it bears the same name.

2. LINEAR OPERATORS IN BANACH SPACE

1°. We call *linear operator* in the Banach space B any endomorphism T: B → B. [As a rule, we shall omit the word "linear" for the sake of brevity. Thus, by "operator" we always mean a bounded (or, equivalently, continuous) linear operator.] The set of all operators in B will be denoted by L(B) (and sometimes by End B). It is an algebra (with identity or unit $E \equiv$ id) under the standard operations of addition, multiplication by a scalar, and multiplication. By a classical theorem of Banach, *an operator* $T \in L(B)$ *is invertible if (and, trivially, only if) it is bijective.* The invertible operators (or, in other words, the automorphisms of the space B), form a multiplicative group Aut B. If $\dim B > 1$ this group is not

Abelian. If $\dim B = n < \infty$, Aut B depends only on n (to within an isomorphism) and is denoted by $GL(n)$ (or, to specify the ground field, by $GL(n, \mathbb{C})$ or $GL(n, \mathbb{R})$). It can be identified with the group of all nonsingular matrices of order n. The set End B of all operators is a multiplicative semigroup. Among its subsemigroups a special interest is attached to the semigroups of powers (iterates) $T^k (k=0,1,2,\ldots)$. For an invertible operator T one can consider the group of its powers T^k $(k=0,\pm1,\pm2,\ldots)$. The study of these objects (and especially of their asymptotic behavior for $k \to \infty$) permits a deep understanding of the nature of the operator T.

The presence of a norm turns $L(B)$ into a Banach space. Moreover, $L(B)$ is a Banach algebra, since the inequality $\|T_1T_2\| \leq \|T_1\|\|T_2\|$ $(T_1, T_2 \in L(B))$ guarantees that multiplication of operators is continuous in the *uniform* (or *norm*, i.e., defined by the norm) operator topology. [Recall that a *Banach algebra* is a Banach space with a continuous, associative, and bilinear multiplication which possesses (unless otherwise stipulated) an identity, and which commutes with multiplication by scalars. If not mentioned otherwise, Banach algebras will be assumed to be complex.] Two other important linear (locally-convex) topologies exist on $L(B)$: the *strong* and the *weak* topologies, defined by the families of seminorms $p_x(T) = \|T_x\|$ $(x \in B)$, and, respectively, $\pi_{x,f}(T) = |f(T_x)|$ $(x \in B,\ f \in B^*)$. In these topologies the multiplication of operators is discontinuous (though it remains continuous in each argument separately). Nevertheless, one has the following simple, yet useful result.

LEMMA. *The multiplication is continuous in the strong topology on every bounded set* $M \subset L(B)$.

PROOF. Suppose $\|T\| \leq c$ for all $T \in M$. Then

$$\|(T_1T_2 - S_1S_2)x\| \leq c\,\|(T_2 - S_2)x\| + \|(T_1 - S_1)S_2 x\|$$

for all $x \in B$ and all $T_1, T_2, S_1, S_2 \in M$. Therefore, if (T_1,T_2) lies in the neighborhood of the pair (S_1,S_2) given by the inequalities $\|(T_1-S_1)S_2x\| < \frac{\varepsilon}{2}$, $\|(T_2-S_2)x\| < \frac{\varepsilon}{2c}$, then T_1T_2 lies in the ε-neighborhood of the operator S_1S_2.

2°. For any operator T in a Banach space B the *conjugate operator* T^* in B^* is defined by the rule $(T^*f)(x) = f(T_x)$ $(f \in B^*, x \in B)$.

THEOREM. *The mapping* $T \to T^*$ *is a norm-preserving Banach space morphism* $L(B) \to L(B^*)$. *Moreover,* $(T_1T_2)^* = T_2^*T_1^*$ *for all* $T_1,T_2 \in L(B)$, $(id_B)^* = id_{B^*}$, *and, if* B is *reflexive, then* $T^{**} = T$ *for all* $T \in L(B)$.

PROOF. The fact that the norm is preserved is seen from the following chain of equalities, which rests on the Hahn-Banach theorem:

$$\|T^*\| = \sup_{\|f\|=1} \|T^*f\| = \sup_{\|f\|=\|x\|=1} |(T^*f)(x)| =$$

$$= \sup_{\|f\|=\|x\|=1} |f(Tx)| = \sup_{\|x\|=1} \|Tx\| = \|T\|.$$

The proof of the remaining assertions is trivial.

□

Let $L \subset B$ be an arbitrary subspace. Its *annihilator* is the subspace $L^\perp = \{f \mid f(x) = 0 \ \forall\, x \in L\}$ of B^*. Similarly (but not completely symmetrically as B is not necessarily reflexive), given an arbitrary subspace $M \subset B^*$ one defines its *annihilator* M as $M^\perp = \{x \mid f(x) = 0, \forall f \in M\} \subset B$. Obviously,

$$L_1 \subset L_2 \Rightarrow L_1^\perp \supset L_2^\perp$$

and

$$M_1 \subset M_2 \Rightarrow M_1^\perp \supset M_2^\perp$$

Also, $(L^\perp)^\perp \supset L$ and $(M^\perp)^\perp \supset M$.

Remark. *Annihilators* can be also defined for arbitrary subsets of B or B^*; they are necessarily subspaces: in fact, *the annihilator of any subset coincides with the annihilator of its closed linear span.*

LEMMA. *The equality* $(L^{\perp})^{\perp} = L$ *holds for every* $L \subset B$. *If* B *is reflexive, then also* $(M^{\perp})^{\perp} = M$ *for every* $M \subset B^*$.

PROOF. The second assertion follows from the first. To prove the first assertion it suffices to observe that, by definition, every linear functional which annihilates the susbspace L annihilates also $(L^{\perp})^{\perp}$.

□

Exercise. *The following "duality formulas" are valid:*

$$(L_1 + L_2)^{\perp} = L_1^{\perp} \cap L_2^{\perp}, \qquad (L_1 \cap L_2)^{\perp} \supset L_1^{\perp} + L_2^{\perp}$$

and

$$(M_1 + M_2)^{\perp} = M_1^{\perp} \cap M_2^{\perp}, \qquad (M_1 \cap M_2)^{\perp} \supset M_1^{\perp} + M_2^{\perp} .$$

If B *is reflexive the inclusions become equalities.*

The indicated rules are well known in the case $\dim B < \infty$ and form the basis of the duality present in projective geometry. We next turn to the operator aspect of this duality.

THEOREM. *Let* B *be a Banach space and* $T \in L(B)$. *Then :*

$$(\operatorname{Im} T)^{\perp} = \operatorname{Ker} T^* , \qquad (\operatorname{Ker} T)^{\perp} \supset \overline{\operatorname{Im} T^*} ,$$

and

$$(\operatorname{Im} T^*)^{\perp} \supset \operatorname{Ker} T , \qquad (\operatorname{Ker} T^*)^{\perp} = \overline{\operatorname{Im} T} .$$

If B *is reflexive these inclusions become equalities.*

PROOF. The equality $T^*f = 0$ is obviously equivalent to $f(Tx) = 0$ $(x \in B)$, i.e., $\operatorname{Ker} T^* = (\operatorname{Im} T)$. It follows that $(\operatorname{Ker} T^*)^{\perp} = (\operatorname{Im} T)^{\perp\perp} = (\overline{\operatorname{Im} T})^{\perp\perp} = \overline{\operatorname{Im} T}$. Let us prove the inclusion $\overline{\operatorname{Im} T^*} \subset (\operatorname{Ker} T)^{\perp}$. Let $f \in \operatorname{Im} T^*$, i.e., $f = T^*g$ for a

$g \in B^*$. Then $f(x) = g(Tx)$ $(\forall\ x \in B)$, and $Tx = 0$ implies $f(x) = 0$. Therefore, $\operatorname{Im} T^* \subset (\operatorname{Ker} T)^{\perp}$, and since the annihilator $(\operatorname{Ker} T)^{\perp}$ is closed, we have also that $\overline{\operatorname{Im} T^*} \subset (\operatorname{Ker} T)^{\perp}$. The inclusion $\operatorname{Ker} T \subset (\operatorname{Im} T^*)^{\perp}$ follows from the preceding one :

$$\operatorname{Ker} T = (\operatorname{Ker} T)^{\perp\perp} \subset (\overline{\operatorname{Im} T^*})^{\perp} = (\operatorname{Im} T^*)^{\perp} .$$

If B is reflexive, then $T^{**} = T$ and hence, by the foregoing proof,

$$\overline{\operatorname{Im} T^*} = (\operatorname{Ker} T^{**})^{\perp} = (\operatorname{Ker} T)^{\perp}$$

and

$$\operatorname{Ker} T = \operatorname{Ker} T^{**} = (\operatorname{Im} T^*)^{\perp}.$$

□

Let us apply this duality between images and kernels in testing the invertibility of the operator T^*.

THEOREM. *In order for* T^* *to be invertible it is necessary and sufficient that* T *be invertible, in which case* $(T^*)^{-1} = (T^{-1})^*$.

PROOF. NECESSITY. If $\operatorname{Im} T^* = B$ and $\operatorname{Ker} T^* = 0$, then $\operatorname{Ker} T = 0$ and $\overline{\operatorname{Im} T} = B$. It suffices to show that $\inf_{\|x\|=1} \|Tx\| > 0$. In fact, in this case, if $y = \lim_{k\to\infty} Tx_k$, then there exists $x = \lim_{k\to\infty} x_k$, and $y = Tx$; hence, $\operatorname{Im} T$ is closed, and so $\operatorname{Im} T = B$. Thus, suppose that $\lim_{k\to\infty} Tz_k = 0$ and $\|z_k\| = 1$. Choose $f_k \in B^*$ such that $f_k(z_k) = 1$, $\|f_k\| = 1$, and set $g_k = (T^*)^{-1} f_k$. Since the sequence $\{g_k\}$ is bounded, we have $\lim_{k\to\infty} g_k(Tz_k) = 0$, whence $\lim_{k\to\infty} f_k(z_k) = 0$, which contradicts the choice of the sequence $\{f_k\}$.

SUFFICIENCY. It follows from the equalities $TT^{-1} = \mathrm{id}$ and $T^{-1}T = \mathrm{id}$ that $(T^{-1})^*T^* = \mathrm{id}$ and $T^*(T^{-1})^* = \mathrm{id}$. Therefore, T^* is invertible and $(T^*)^{-1} = (T^{-1})^*$.

□

Let $B = H$ be a Hilbert space. Then, by Riesz's theorem on the general form of linear functionals on H : $f(x) = (x, y_f)$, we can identify $H^* = H$ (recall, however, that the mapping $f \to y_f$ is anti-isometric, i.e., it enjoys all the properties of an isometry except for homogeneity, which is replaced by $y_{\alpha f} = \bar{\alpha} y_f$). Accordingly, we can regard T^* as acting in the space H itself, so that $(Tx,y) = (x,T^*y)$. In this context we are naturally led to the notion of a *self-adjoint operator* $T \in L(H)$, i.e., a T such that $T^* = T$.

Exercise. *The self-adjoint operators form a real subspace of the Banach space* $L(H)$.

With each self-adjoint operator T we can associate the quadratic form or functional (Tx,x), $x \in H$.

Exercise. $(Tx,x) \equiv 0$ *implies* $T = 0$.

A self-adjoint operator T is said to be *nonnegative* (written $T \geqslant 0$) if $(Tx,x) \geqslant 0$ for all x, *positive* (written $T > 0$) if $(Tx,x) > 0$ for all $x \neq 0$, and *positive definite* (written $T \gg 0$) if there exists a constant $c > 0$ such that $(Tx,x) \geqslant c\|x\|^2$ for all x.

Exercise. *The operator* A^*A *is self-adjoint and nonnegative for every* $A \in L(H)$. *Moreover,*

$$A^*A > 0 \Leftrightarrow \operatorname{Ker} A = 0$$

and

$$A^*A \gg 0 \Leftrightarrow A \text{ is invertible.}$$

A detailed exposition of the theory of self-adjoint operators is given in the monograph of N. I. Akhiezer and I. M. Glazman [1]. The importance of this theory is explained, among others, by its connection with quantum mechanics. In point of fact, this connection has exerted a considerable influence on the development of the theory itself (see J. von Neumann's book [36], published in 1936).

3°. From an algebraic viewpoint the simplest operators are the *idempotents*, i.e., the operators T such that $T^2 = T$. The geometrical equivalent of this property is that T is a *projection*, that is, there is a subspace $L \subset B$ such that $Tx \in L$ for all x and $Tx = x$ whenever $x \in L$ (in this case we say that T is a *projection onto* L). In fact, it follows from these conditions that $T^2 = T$; conversely, if $T^2 = T$, then we put $L = \operatorname{Im} T$.

Exercise. *If* T *is a projection, then* $B = \operatorname{Im} T \dotplus \operatorname{Ker} T$ *and* $T|\operatorname{Im} T = \mathrm{id}$ *(here* $\dotplus$ *stands for direct sum). Conversely, if* $B = L \dotplus N$, *with* L *and* N *subspaces, then the operator* T *specified by the conditions* $T|L = \mathrm{id}$, $T|N = 0$, *is a projection and* $\operatorname{Im} T = L$, $\operatorname{Ker} T = N$.

Projections are customarily denoted by the letter P. Let $\{P_\lambda\}$ be a (possibly infinite) family of projections in B. $\{P_\lambda\}$ is called a *resolution of identity* if it is

1) *complete*, i.e., $\overline{\sum_\lambda \operatorname{Im} P_\lambda} = B$,

2) *total*, i.e., $\bigcap_\lambda \operatorname{Ker} P_\lambda = 0$,

and

3) *algebraically orthogonal*, i.e., $P_\lambda P_\mu = 0$ for all $\lambda \neq \mu$.

Example. Let $\mathbb{T}$ denote the unit circle and $L_1(\mathbb{T})$ the space of scalar functions on $\mathbb{T}$ which are summable with respect to the Lebesgue measure. Let $c_n(\phi)$ denote the n-th Fourier coefficient of the function $\phi \in L_1(\mathbb{T})$:

$$c_n(\phi) = \frac{1}{2\pi}\int_{-\pi}^{\pi} \phi(s)e^{-ins}ds, \quad n = 0, \pm 1, \pm 2, \ldots .$$

The operators P_n given by $(P_n\phi)(t) = c_n(\phi)e^{int}$ are projections in $L_1(\mathbb{T})$ and $\{P_n\}$ is a resolution of identity. The classical Fourier series of ϕ is written

$$\sum_{n=-\infty}^{\infty} P_n\phi .$$

Guided by this example we give the following general definition. Let $\{P_\lambda\}$ be a resolution of identity in the Banach space B. The *Fourier series* of the vector $x \in B$ (with respect to

$\{P_\lambda\}$) is the formal (generally speaking, unordered) series $\sum_\lambda P_\lambda x$ (as in the classical case, we write $x \sim \sum_\lambda P_\lambda x$). We shall not dwell further upon the notion of convergence for such general Fourier series, which needs to be made precise.

Exercise. *Suppose the Fourier series of two vectors* $x,y \in B$ *coincide. Then* $x = y$ *("uniqueness theorem").*

Given an arbitrary resolution of identity $\{P_\lambda\}$, we say that B is the *topological direct sum* of the subspaces $L_\lambda = \operatorname{Im} P_\lambda$. The family $\{L_\lambda\}$ is complete (i.e., $\sum_\lambda L_\lambda = B$) and minimal (i.e., for every λ, $[\sum_{\mu\neq\lambda} L_\lambda] \cap L_\mu = 0$).

Obviously, $\|P\| \leqslant 1$ for every projection $P \neq 0$. If, in addition, $\|P\| = 1$, then P is called an *orthogonal projection* or *orthoprojector*. A resolution of identity is said to be *orthogonal* if it consists of orthogonal projections (in which case we speak about an *orthogonal sum of subspaces*). Such is, for example, the classical Fourier resolution in $L_1(\mathbb{T})$ (and also in any space $L_p(\mathbb{T})$ with $1 \leqslant p \leqslant \infty$). If P is an orthogonal projection in B, the resolution of identity $\{P, E-P\}$ is termed *semiorthogonal*. A subspace $L \subset B$ for which there exists an orthogonal projection P such that $\operatorname{Im} P = L$ is said to *admit an orthogonal complement* (or to be *orthocomplemented*).

Exercise 1. *Let* P *be a projection in a Hilbert space* H. *The following assertions are equivalent :*

1) P *is an orthogonal projection ;*
2) *the subspaces* $\operatorname{Im} P$ *and* $\operatorname{Ker} P$ *are orthogonal ;*
3) $P^* = P$.

Exercise 2. *If* P *is an orthogonal projection in a Hilbert space, then so is* $E - P$.

In an arbitrary Banach space this is false though, of course, $E - P$ is a projection for every projection P.

4°. An operator T in the Banach space B is called a *contraction* if $\|T\| \leq 1$ (i.e., $\|Tx\| \leq \|x\|$ for all $x \in B$). The set of all contractions is a subsemigroup of the multiplicative semigroup $\text{End } B$. The operator T is called an *isometry* if it preserves the norm (i.e., $\|Tx\| = \|x\|$ for all $x \in B$) and is invertible. The isometries form a subgroup $\text{Iso } B$ of $\text{Aut } B$.

Exercise. $\text{Iso } B$ *coincides with the group of invertible elements in the semigroup of all contractions.*

In a Hilbert space H isometries are called *unitary operators*, and the group $\text{Iso } H$ is alternatively denoted by $U(H)$ (or, in the real case, by $O(H)$) and is called the *unitary* (respectively, *orthogonal*) group of H. It depends on $\dim H$ only (to within an isomorphism ; here dimension is understood as the cardinality of an orthonormal basis). If $\dim H = n$ one writes $U(n)$ and one usually identifies $U(n)$ with the group of unitary matrices of order n.

Exercise. $T \in U(H) \Leftrightarrow T^*T = TT^* = E$.

If T *is a contraction in the Banach space* B, *then obviously the semigroup of powers* T^k, $k \geq 0$, *is bounded.*

Exercise. *Suppose the semigroup of powers* T^k *is bounded. Then the norm* $\|x\|_T = \sup_{k\geq 0}\|T^k x\|$ *is equivalent to the original norm on* B, *and, relative to* $\|\cdot\|_T$, T *is a contraction.*

The asymptotic behavior of a bounded semigroup $\{T^k\}_{k\geq 0}$ is described, to a certain extent, by the so-called *statistical ergodic theorem*. This theorem (J. von Neumann, 1932) emerged from an analysis of the foundations of statistical mechanics and gave impetus to the development of a vast mathematical discipline, ergodic theory (a concise yet brilliant introduction to this theory is given in P. Halmos' book [18]; deep modern results are discussed in the monograph of I. P. Cornfeld, Ya. G. Sinai, and S. V. Fomin [9]). We give here the following generalization of the statistical ergodic theorem, due to E. Lorch (1939).

THEOREM. *Suppose the semigroup* $\{T^k\}_{k\geq 0}$ *is bounded. Then the strong limit*

$$\boxed{P_1 = \lim_{n\to\infty} \frac{1}{n} \sum_{k=0}^{n} T^k}$$

exists and is a projection onto the subspace of fixed points of the operator T. *If* T *is a contraction, then* P_1 *is an orthogonal projection.*

This result is a consequence of the following assertion, valid in an arbitrary Banach space.

THEOREM. *Suppose* $\sup_{k\geq 0} \|T^k\| < \infty$. *Set* $R = T - E$, $T_n = \sum_{k=0}^{n} T^k$, *and let* M_T *denote the subspace of all vectors* $x \in B$ *for which the limit* $P_1x = \lim_{n\to\infty}\frac{1}{n} T_n x$ *exists. Then* $M_T = \operatorname{Ker} R \dotplus \overline{\operatorname{Im} R}$. *Moreover,* $P_1|\operatorname{Ker} R = \mathrm{id}$ *and* $P_1|\overline{\operatorname{Im} R} = 0$.

PROOF. Let $x \in \operatorname{Im} R$, $x = Ty - y$. Then $\frac{1}{n} T_n x = \frac{1}{n}(T^n y - y) \to 0$ as $n \to \infty$. Consequently, $\operatorname{Im} R \subset M_T$, whence $\overline{\operatorname{Im} R} \subset M_T$. If $x \in \operatorname{Ker} R$, then $Tx = x$, whence $\frac{1}{n} T_n x = x$. We see that $\operatorname{Ker} R \subset M_T$. Therefore, $\operatorname{Ker} R \dotplus \overline{\operatorname{Im} R} \subset M_T$. The subspaces $\operatorname{Ker} R$ and $\overline{\operatorname{Im} R}$ are independent, i.e., $\operatorname{Ker} R \cap \overline{\operatorname{Im} R} = 0$, for $\frac{1}{n} T_n|\operatorname{Ker} R = \mathrm{id}$ and $\frac{1}{n} T_n|\overline{\operatorname{Im} R} \to 0$ as $n \to \infty$. Finally, let $x \in M_T$, $y = \lim_{n\to\infty}\frac{1}{n} T_n x$, and $z = x - y$. Then $Ry = \lim_{n\to\infty}\frac{1}{n}(T^n x - x)$, i.e., $y \in \operatorname{Ker} R$. At the same time, $z = \lim_{n\to\infty}\frac{1}{n} \sum_{k=0}^{n-1}(x - T^k x)$, and $x - T^k x = - RT_k x$. This shows that $z \in \overline{\operatorname{Im} R}$. Consequently, $x = y + z$ $\operatorname{Ker} R \dotplus \overline{\operatorname{Im} R}$.

□

To prove the von Neumann-Lorch theorem it remains only to check that $M_T = B$ whenever the space B is reflexive. But $M_T^{\perp} = (\operatorname{Ker} R)^{\perp} \cap (\overline{\operatorname{Im} R})^{\perp} = \overline{\operatorname{Im} R^*} \cap \operatorname{Ker} R^*$, and this last intersection reduces to zero for the same reason as for R : $R^* = T^* - E$ and $\sup_{k\geq 0} \|(T^*)^k\| < \infty$, because $\|(T^*)^k\| = \|T^k\|$. □

3. SPECTRUM AND RESOLVENT OF LINEAR OPERATORS

1°. Let T be an operator in the Banach space B. The complex number λ is called a *regular value* (or *regular point*) of T if $T - \lambda E$ is an invertible operator. In this case the inverse operator $R_\lambda = (T - \lambda E)^{-1}$ is called the *resolvent* of T (at the point λ). The set $\text{reg}\, T$ of regular values of the operator T is called its *domain of regularity* or *resolvent set*. The complement $\mathbb{C} \setminus \text{reg}\, T = \text{spec}\, T$ is called the *spectrum* of T. It is with this fundamental definition that the spectral theory of linear operators, built in the XXth century through efforts of I. Fredholm, D. Hilbert, J. von Neumann, F. Riesz, and other prominent mathematicians, starts.

Inside the spectrum $\text{spec}\, T$ of an arbitrary operator T we distinguish the so-called *discrete spectrum* $\text{spec}_d T$, which consists of all eigenvalues of T. As in linear algebra, the complex number λ is called an *eigenvalue* of T if the operator $T - \lambda E$ is not injective. In this case the subspace $\text{Ker}(T - \lambda E)$ is called an *eigensubspace* of T, and its nonzero elements (i.e., the nontrivial solutions of the equation $Tx = \lambda x$) are called the *eigenvectors* of T corresponding to λ.

Example. In a finite dimensional space B every injective linear operator T is nonsingular (i.e., $\det T \neq 0$), and hence surjective. Therefore, *in a finite dimensional space the whole spectrum of any operator consists of eigenvalues*. In addition, it is finite : $|\text{Spec}\, T| < \dim B$ (here $|\ |$ stands for cardinality).

This is in contrast to the following example.

Example. In $C[0,1]$ consider the operator T of multiplication by the independent variable : $(T\phi)(t) = t\phi(t)$. Its spectrum is the segment $[0,1]$. In fact, the equation $t\phi(t) - \lambda\phi(t) = \psi(t)$ has a solution in $C[0,1]$ for every $\psi \in C[0,1]$ if and only if $\lambda \notin [0,1]$. At the same time, the equation $t\theta(t) = \lambda\theta(t)$ has only the trivial solution.

The discrete spectrum $\mathrm{spec}_d T$ is a subset of the wider *approximate spectrum* $\mathrm{spec}_a T$ of T, which consists of the quasi-eigenvalues of T. The complex number λ is called a *quasi-eigenvalue* of the operator T if there exists a sequence of vectors $\{x_k\}_{k=1}^{\infty}$ which does not converge to zero, such that $Tx_k - x_k \to 0$ (alternative terms are *approximate eigenvalue* and *almost-eigenvalue*). A sequence with the indicated property is also referred to as a *quasi-eigensequence*, and one can always assume that it is normalized so that $\|x_k\| = 1$ for all k. In the previous example the entire spectrum is approximate. In fact, if $\lambda \in [0,1]$, then the functions $\max(1 - k|t-\lambda|, 0)$ form a quasi-eigensequence.

The difference $\mathrm{spec}\, T \setminus \mathrm{spec}_a T$ is called the *residual spectrum* of T, denoted $\mathrm{spec}_r T$.

We next discuss the simplest topological properties of the spectrum.

THEOREM. *The domain of regularity* $\mathrm{reg}\, T$ *of any operator* T *is open. The resolvent* R_λ *of* T *is holomorphic in* $\mathrm{reg}\, T$.

PROOF. Let $\lambda_0 \in \mathrm{reg}\, T$. Consider the power series $S(\lambda) = \sum_{k=0}^{\infty} R_{\lambda_0}^{k+1}(\lambda - \lambda_0)^k$. It manifestly converges in the disk $|\lambda - \lambda_0| < \|R_{\lambda_0}\|^{-1}$. Its sum is holomorphic in the disk of convergence. At the same time, $S(\lambda)(T - \lambda E) = (T - \lambda E)S(\lambda) = E$, as is readily checked. Therefore, the disk of convergence, of radius $\rho \geq \|R_{\lambda_0}\|^{-1} > 0$, is contained in $\mathrm{reg}\, T$, and $S(\lambda) = R_\lambda$ at any point λ of this disk.

□

COROLLARY. *The spectrum of any operator is closed.*

□

However, to get a meaningful spectral theory we need the following important result.

THEOREM. *The spectrum of any operator is not empty.*

The proof relies on the following

LEMMA. *Let* T *be an arbitrary operator. Then every value* λ *such that* $|\lambda| > \|T\|$ *is regular, and in this domain* $\|R_\lambda\| \leqslant (|\lambda| - \|T\|)^{-1}$.

PROOF. This follows from the convergence of the Laurent series $\sum_{k=0}^{\infty} T^k \lambda^{-k+1}$ for $|\lambda| > \|T\|$. Its sum coincides with R_λ in the domain of convergence, which immediately yields the desired estimate.

□

Assuming now that the spectrum of the operator T is empty, the resolvent R_λ of T is an entire operator-valued function which tends to zero as $\lambda \to \infty$. By Liouville's theorem, $R_\lambda = 0$, which is impossible since R_λ is invertible. This proves that spec $T \neq \emptyset$.

□

Annother consequence of the preceding lemma is that spec T *is contained in the disk* $|\lambda| \leqslant \|T\|$. Therefore, *the spectrum of* T *is compact.* The smallest number $\rho > 0$ such that spec T is contained in the disk $|\lambda| \leqslant \rho$ is called the *spectral radius* of the operator T, denoted $\rho(T)$. By the foregoing discussion, $\rho(T)$ $\|T\|$. This observation can be completed to the following exact *formula of Gelfand* :

$$\boxed{\rho(T) = \lim_{k\to\infty} \|T^k\|^{1/k} .} \tag{1}$$

The existence of the limit in (1) is guaranteed by a known *theorem of Fekete* asserting that if $\{a_k\}_{k=1}^{\infty}$ is a subadditive sequence of real numbers, i.e., $a_{k+j} \leqslant a_k + a_j$, then $k^{-1}a_k$ converges to its infimum. In particular, we can take $a_k = \ln \|T^k\|$ to conclude that the limit in (1) exists and equals $\inf_k \|T^k\|^{1/k}$. It remains to prove equality (1). To this end we use the Laurent expansion of the resolvent. Its outer radius of convergence is equal, by the Cauchy-Hadamard formula, to the limit (1). On the other hand, it is $\geqslant \rho(T)$, because the domain of convergence of the series is contained in reg T. At the same time, it is $\leqslant \rho(T)$. because the Laurent series of the holomorphic function R_λ converges in the domain $|\lambda| > \rho(T)$.

Exercise 1. *Let* $b_k > 0$, $k \geqslant 1$, *be a sequence of numbers satisfying* $b_{k+j} \leqslant b_k b_j$. *Then there exists an operator* T *in a Hilbert space such that* $\|T^k\| = b_k$ *for all* k.

Exercise 2. *If the semigroup* $\{T^k\}$ *of powers of* T *is bounded, then* $\rho(T) \leqslant 1$.

Exercise 3. *If* $\rho(T) < 1$, *then the semigroup* $\{T^k\}$ *is bounded (and even* $\|T^k\| \to 0$).

Exercise 4. *For every operator* T *in the Banach space* B *and every* $\varepsilon > 0$ *there in an equivalent norm* $\|\cdot\|'$ *such that* $\|T\|' < \rho(T) + \varepsilon$. *Therefore,* $\rho(T) = \inf_\nu \nu(T)$, *where* ν *runs through the set of norms on* L(B) *corresponding to all norms on* B *equaivalent to the given one. If* B *is a Hilbert space one can confine ourselves to equivalent Hilbert norms.*

The circle $|\lambda| = \rho(T)$ obviously contains at least one point of the spectrum of T.

THEOREM. *Every point of the spectrum lying on the circle* $|\lambda| = \rho(T)$ *belongs to the approximate spectrum of* T.

This is a straightforward consequence of the following

THEOREM. *The residual spectrum of any operator is open.*

PROOF. We first remark that every point λ in the residual spectrum is *quasi-regular*, meaning that there is a constant $c > 0$ such that $\|Tx - \lambda x\| \geqslant c\|x\|$ for all x. Next, for each quasi-regular point λ the operator $T - \lambda E$ is injective and $\mathrm{Im}(T - \lambda E)$ is closed. Now let $\lambda \in \mathrm{spec}_r T$. Then $\mathrm{Im}(T - \lambda E) \neq B$: otherwise, λ would be a regular point by the foregoing argument. Hence, there exists a linear functional $f \neq 0$ which annihilates $\mathrm{Im}(T - \lambda E)$. Consider a disk $|\lambda - \mu| < \varepsilon$, where $\varepsilon < \frac{1}{2}c$. It consists of quasi-regular points : $\|Tx - \mu x\| \geqslant (c - \varepsilon)\|x\|$. Let us show that the points of this disk belong to the residual spectrum of T. Suppose this is not true for the point μ. Then $\mu \in \text{reg } T$ and $\|R_\mu\| \leq (c - \varepsilon)^{-1}$. Consider the operator $(T - \lambda E)R_\mu = E + (\lambda - \mu)R_\mu$. The functional f annihilates its image, and so $f(x) = (\lambda - \mu)f(R_\mu x)$ for all $x \in B$. But then $|f(x)| \leq \varepsilon(c - \varepsilon)^{-1}\|f\|\|x\|$, which is impossible if $f \neq 0$, because $\varepsilon(c - \varepsilon)^{-1} < 1$.

□

If now $\lambda \in \text{spec } T$ and $|\lambda| = \rho(T)$, then $\lambda \in \text{spec}_a T$: otherwise, $\lambda \in \text{spec}_r T$ and then $\mu \in \text{spec}_r T$ for all μ close to λ, which is impossible if $|\mu| > \rho(T)$.

The same arguments show that *the topological boundary of the spectrum of any operator* T *is contained in the approximate spectrum* $\text{spec}_a T$.

Exercise. *If* $T \in \text{Iso } B$, *then* $\text{spec } T$ *lies on the unit circle and is equal to* $\text{spec}_a T$.

2°. There is a class of operators, the compact ones, for which the spectrum has a particularly simple structure. [Recall that an operator T is said to be *compact* if the image of the unit ball under T is precompact. Compact operators for a closed two-sided ideal in the Banach algebra $L(B)$.]

THEOREM. *The spectrum of a compact operator* T *in an infinite dimensional space* B *consists of the point* 0 *and an at most countable (possibly empty) set of nonzero eigenvalues. The unique limit point of this set, in case it is infinite, is* 0.

PROOF. Let $\lambda \in \text{spec}_a T$, $\lambda \neq 0$, and let $\{x_k\}$ be a corresponding quasi-eigensequence ($\|x_k\| = 1$). Since T is compact, we can assume that $\{Tx_k\}$ converges. Also, since $Tx_k - \lambda x_k \to 0$ and $\lambda \neq 0$, the limit $x = \lim_{k\to\infty} x_k$ exists, and $\|x\| = 1$, $Tx = \lambda x$. Therefore, every point $\lambda \neq 0$ in $\text{spec}_a T$ is an eigenvalue. We show that for every $\delta > 0$ the domain $|\lambda| > \delta$ contains only a finite number of eigenvalues. Assuming that this is not so, let λ_k, $k = 1,2,3,\ldots$ be distinct eigenvalues such that $|\lambda_k| > \delta$, and let x_k be corresponding eigenvectors. Consider the increasing sequence of subspaces $L_n = \mathrm{Lin}(x_1,\ldots,x_n)$, $n \geq 1$ (where Lin stands for "linear span"). Choose a vector $y_n \in L_n$ such that $\|y_n\| = 1$ and $\mathrm{dist}(y_n, L_{n-1}) > \frac{1}{2}$. Write y_n as a linear combination $y_n = \sum_{k=1}^{n} \alpha_{kn} x_k$. Then $Ty_n = \sum_{k=1}^{n} \alpha_{kn}\lambda_k x_k$, and so $Ty_n - \lambda_n y_n \in L_{n-1}$. Consequently, $\mathrm{dist}(Ty_n, L_{n-1}) =$ $= |\lambda_n| \mathrm{dist}(y, L_{n-1}) > \frac{\delta}{2}$. Since $Ty_m \in L_{n-1}$ for any $m < n$, it follows that $\|Ty_n - Ty_m\| > \frac{\delta}{2}$ if $m \neq n$. This contradicts the compactness of T, which guarantees that the sequence $\{Ty_k\}$ contains a convergent subsequence.

Thus, the set of nonzero eigenvalues of T is at most countable, and its unique possible limit point is $\lambda = 0$. This point always belongs to spec T, as the existence of T^{-1} would imply the compactness of the identity operator E, which is impossible if $\dim B = \infty$.

It remains to notice that a compact operator has no residual spectrum. In fact, the residual spectrum is open, and its topological boundary belongs to the spectrum, and so is contained in the approximate spectrum. It follows that the boundary of $\mathrm{spec}_r T$ is at most countable, which is not possible for a bounded open set.

□

COROLLARY. *The eigensubspaces of a compact operator corresponding to nonzero eigenvalues are finite dimensional.*

In fact, in each such subspace the spectrum of the operator does not contain the point $\lambda = 0$.

□

3°. What can be said about the spectrum of a self-adjoint operator in a Hilbert space H? The answer follows from general considerations.

THEOREM. *Let* T *be an operator in a Banach space* B. *Then* spec $T^* =$ spec T.

PROOF. $(T - \lambda E)^* = T^* - \lambda E$ and so the operator $T^* - \lambda E$ is invertible if and only if $T - \lambda E$ is invertible.

□

<u>Exercise.</u> $\operatorname{spec}_r T \subset \operatorname{spec}_d T^*$.

As we know, in the Hilbert space case T^* is defined intrinsically (if we canonically identify H^* and H). Since $(T - \lambda E)^* = T^* - \bar{\lambda}E$, it follows that *in Hilbert space* $\operatorname{spec} T^* = (\operatorname{spec} T)^*$, where in the right-hand side $*$ stands for complex conjugation. An immediate consequence is the following

THEOREM. *The spectrum of any self-adjoint operator lies on the real axis.*

□

COROLLARY. *A self-adjoint operator has no residual spectrum.*

□

<u>Exercise.</u> *For every compact set* $S \subset \mathbb{R}$ *there is a self-adjoint operator* T *such that* $\operatorname{spec} T = S$.

In the class of self-adjoint operators the relation between spectral radius and norm takes the simplest form.

THEOREM. *Let* T *be a self-adjoint operator. Then* $\rho(T) = ||T||$.

PROOF. Let $\mu = \sup_{||x||=1} |(Tx,x)|$. Obviously, $\mu \leqslant ||T||$. It follows from the identity

$$\operatorname{Re}(Tx,y) = \frac{1}{4}\{(T(x+y),x+y) - (T(x-y),x-y)\}$$

that

$$|\operatorname{Re}(Tx,y)| \leqslant \frac{\mu}{4}\{||x+y||^2 + ||x-y||^2\} = \frac{\mu}{2}\{||x||^2 + ||y||^2\}.$$

Consequently, $|(Tx,y)| \leqslant \mu$ for all x,y such that $||x|| = ||y|| = 1$, whence $||T|| \leqslant \mu$. Thus, $||T|| = \mu$. We may assume, with no loss of generality, that $\mu = \sup_{||x||=1} (Tx,x)$. Pick a sequence of unit-norm vectors $\{x_k\}$ such that $(Tx_k,x_k) \to \mu$. Then

$$\overline{\lim_{k\to\infty}} ||Tx_k - \mu x_k||^2 \leqslant \overline{\lim_{k\to\infty}} (||Tx_k||^2 - \mu^2) \leqslant 0.$$

Therefore, $Tx_k - \mu x_k \to 0$, i.e., $\mu \in \operatorname{spec} T$. Since $\mu = \|T\|$, we conclude that $\rho(T) = \|T\|$.

□

Exercise 1. *For every operator* T *in Hilbert space* $\|T\| = \{\|T^*T\|\}^{1/2} = \{\rho(T^*T)\}^{1/2}$.

Exercise 2. *Suppose* T *is a self-adjoint operator and* $\operatorname{spec} T = \{0\}$. *Then* $T = 0$.

Exercise 3. *The spectrum of any nonnegative self-adjoint operator* T *lies on the half-line* $\lambda \geq 0$.

We now know enough about self-adjoint and compact operators to give a full decription of the structure of compact self-adjoint operators.

THE SPECTRAL THEOREM. *Let* T *be a compact self-adjoint operator in a Hilbert space* H. *Let* $\{\lambda_k\}$ *be the sequence of all eigenvalues of* T *(including* $\lambda_0 = 0$, *if it is an eigenvalue) and let* $\{H_k\}$ *be the sequence of corresponding eigensubspaces. Then one has the orthogonal decomposition*

$$H = \bigoplus_{k=0}^{\infty} H_k .$$

PROOF. The fact that the subspaces H_k are pairwise orthogonal is established in exactly the same manner as in finite dimensional linear algebra : if $Tx = \lambda_k x$ and $Ty = \lambda_i y$, with $i \neq k$, then $(x,y) = (\lambda_k - \lambda_i)^{-1}\{(Tx,y) - (x,Ty)\} = 0$. Suppose now that the closure of the sum $\sum H_k$ is not equal to H. Then its orthogonal complement $L \neq 0$ and is invariant under T, as in the finite dimensional case. The operator $T|L$ in L is self-adjoint and compact. On the other hand, it has no eigenvectors, since all such vectors lie in L . Consequently, $\operatorname{spec}(T|L) = 0$, whence $T|L = 0$, i.e., $L \subset H_0$, which contradicts the inclusion $H_0 \subset L$. □

COROLLARY. *Every compact self-adjoint operator possesses an orthonormal basis of eigenvectors.*

Producing a basis of eigenvectors for an operator (or of certain analogues of such a basis) is one of the main problems of spectral theory. An extensive literature is devoted to this subject. Its importance resides in both its intrinsic depth and its connections with various applications, in particular, with problems of mathematical physics, from which it in fact emerged (see, for example, the treatise of R. Courant and D. Hilbert [10]). As one of the related formulations we mention the completeness problem for the system of eigenvectors or, equivalently, for the system of eigensubspaces. We remaind the reader that a system of vectors is said to be *complete* if its linear span is dense. For example, according to the Spectral Theorem, *the system of eigenvectors of any compact self-adjoint operator is complete.*

4. INVARIANT SUBSPACES

1°. If the space B is finite dimensional and $\dim B > 1$, every operator $T \in L(B)$ has a nontrivial invariant subspace. Such is, for example, the one-dimensional subspace spanned by any eigenvector of T. The question of whether any operator in an infinite dimensional Banach space admits a nontrivial invariant subspace is still open, despite the attempts of numerous authors (a counterexample is constructed in the work of C. J. Read, *Bull. London Math. Soc.*, 16 (1984), 337-401). For compact operators an affirmative answer was given by N. Aronszajn and K. Smith (1954), whose work opens by the Hilbert space case proof, found back in 1935 by J. von Neumann. A. Bernstein and A. Robinson (1966) replaced the requirement that the operator T be compact by the compactness of some polynomial $p(T) \neq 0$. Their proof relies on "nonstandard analysis", a branch of mathematical logic (more precisely, of model theory), based on nonstandard interpretations of number systems. However, immediately after, P. Halmos found a "standard" proof of the Bernstein-Robinson theorem. After that no progress was made in the general problem of existence of invariant subspaces until 1973, when V. I. Lomonosov proposed a entirely new approach which allowed him to obtain the following remarkable result.

THEOREM 1. *Suppose that there is a compact operator* $V \neq 0$ *which commutes with* T. *Then* T *has a nontrivial invariant subspace.*

□

In point of fact, Lomonosov proved the following stronger result.

THEOREM 2. *Let* $\dim B = \infty$. *If* V *is a compact operator in* B *and* $V \neq 0$, *then the set of all operators that commute with* V *possesses a common nontrivial invariant subspace.*

□

Generally, a set $M \subset L(B)$ of operators is said to be *reducible* if there exists a nontrivial subspace invariant under M, i.e., under every operator $T \in M$. [The notion of reducibility (and especially the opposite notion of *irreducibility*) plays a basic role in representation theory, and in fact it emerged with the theory itself.] We also mention that for every set $M \subset L(B)$ its centralizer (or commutant) $M' = \{T \mid T \in L(B),\ TA = AT\ \forall A \in M\}$ is a closed subalgebra of $L(B)$. In this language Theorem 2 can be restated as follows : *the centralizer of any compact operator* $V \neq 0$ *is reducible (assuming* $\dim B = \infty$ *).* The key ingredient in the proof of this theorem given by Lomonosov is the following Lemma, which is also important in its own right.

LEMMA. *Let* $A \subset L(B)$ *be an irreducible subalgebra which contains a compact operator* $V \neq 0$ *and the identity operator* E. *Then there exists a compact operator* $U \in A$ *which has a fixed point* $\bar{x} \neq 0$.

PROOF. Consider for each vector $y \in B$ its *orbit* $O(y) = \{z \mid z = Ay,\ A \in A\}$. It is the smallest linear manifold in B which contains y and is invariant under all operators $A \in A$. Its closure $\overline{O(y)}$ is thus the smallest A-invariant subspace containing y. If $y \neq 0$, then $\overline{O(y)} \neq 0$, and since the algebra A is irreducible, $\overline{O(y)} = B$. Hence, the orbit of every vector $y \neq 0$

is dense. Pick an arbitrary vector x_0 such that $Vx_0 \neq 0$ and set $\alpha = 2\|V\|/\|Vx_0\|$, $x_1 = \alpha x_0$. Consider the ball $Q = \{x \mid \|x - x_1\| \leq 1\}$. If $x \in Q$, then $\|Vx\| \geq \|Vx_1\| - \|V\| = \alpha\|Vx_0\| - \|V\| = \|V\| > 0$. The closure $K = \overline{VQ}$ of the image of Q under V is compact thanks to the compactness of the operator V. Choose for each $y \in K$ an operator $A_y \in A$ such that $\|A_y y - x_1\| < 1$. Then the inequality $\|A_y v - x_1\| < 1$ holds for all vectors v in a neighborhood of y. Since K is compact, there is a finite family $\{A_1, \ldots, A_n\} \subset A$ such that at each point $v \in K$ at least one of the inequalities $\|A_i v - x_1\| < 1$, $1 \leq i \leq n$, is satisfied. Now define the function $\phi(t) = \max(1-t, 0)$ on the half-line $[0, \infty)$ and consider the map $\Phi: K \to B$ given by the rule

$$\Phi v = \sum_{i=1}^{n} \phi_i(v) A_i v ,$$

where

$$\phi_i(v) = \frac{\phi(\|A_i v - x_1\|)}{\sum_{j=1}^{n} \phi(\|A_j v - x_1\|)} .$$

It is obvious that the coefficients $\phi_i(v)$ are defined, continuous, and nonnegative on the compact set K, and satisfy $\sum_{i=1}^{n} \phi_i(v) = 1$. It follows that the map Φ is continuous, and hence that its image ΦK is compact. Furthermore, ΦK is contained in the original ball Q: indeed, if $A_i v \notin Q$ for some v, i.e., $\|A_i v - x_1\| > 1$, then $\phi_i(v) = 0$. Finally, consider the map $\Psi = \Phi V : Q \to Q$. It is continuous and has compact image. By Schauder's principle, Ψ has a fixed point $\bar{x}$, i.e., $\sum_{i=1}^{n} \phi_i(V\bar{x}) A_i V\bar{x} = \bar{x}$. Therefore, $\bar{x}$ is a fixed point of the compact operator $U = \sum_{i=1}^{n} \phi_i(V\bar{x}) A_i V \in A$. □

COROLLARY. *Let* $A \subset L(B)$ *be an irreducible subalgebra which contains a compact operator* $V \neq 0$, *and let* A' *denote the centralizer of* A. *Then every* $T \in A'$ *is a scalar operator, i.e.,* $T = \lambda E$, *where* λ *is a scalar.*

This is a far-reaching generalization of the classical Schur Lemma, which is concerned with the case $\dim B < \infty$.

PROOF. We may assume, with no loss of generality, that $E \in A$. Lomonosov's lemma yields a compact operator $U \in A$ which has a fixed point. Let F denote the finite dimensional subspace of all fixed points of U ; $F \neq 0$. Then F is T-invariant, since $TU = UT$. It follows that T has an eigenvector in F. Let λ be the corresponding eigenvalue, and let L be the eigensubspace of T associated with λ in the full space B. Since, by the preceding argument, $L \neq 0$, and since L is invariant under all operators $A \in A$ (because $AT = TA$, $\forall A \in A$), the irreducibility of A implies that $L = B$. Therefore, $T = \lambda E$.

□

Now Theorem 2 (and consequently Theorem 1) can be proved in a few words : if, under its hypotheses, the centralizer $\{V\}'$ is irreducible, then the fact that it is a subalgebra and contains $V \neq 0$ implies that $V = \lambda E$. Since $\dim B = \infty$, this contradicts the compactness of the operator $V \neq 0$.

□

We mention that the results of Lomonosov have generated a series of investigations that continues to this day.

2°. Schur's Lemma alluded to above asserts that *the centralizer* A' *of any irreducible subalgebra* A *of the algebra of endomorphisms of a finite dimensional vector space consists of scalar operators.* The proof is very simple and goes like this : Suppose $T \in A'$ is not scalar. Then T possesses a proper invariant subspace, and we thus obtain a proper subspace invariant under A. An example of irreducible subalgebra of $L(B)$ is $L(B)$ itself. A very important fact is that in the finite dimensional case this is the only example possible.

BURNSIDE'S THEOREM. *Suppose* $\dim B = n < \infty$ *and let* $A \subset L(B)$ *be an irreducible subalgebra. Then* $A = L(B)$.

PROOF. Since B is irreducible, $Ay = B$ for every vector $y \neq 0$ (all orbits are closed because $\dim B < \infty$). For each $k = 1,\dots n$ consider the direct sum $B^k = B \dotplus \dots \dotplus B$ of k

copies of B. Pick a basis $e_1,\dots,e_n$ in B and denote by $\sigma_k : B \to B^k$ the linear map which sends each operator $A \in A$ into the element $Ae_1 \dotplus \dots \dotplus Ae_k \in B^k$. Clearly, $\operatorname{Ker}\sigma_1 \supset \dots \supset \operatorname{Ker}\sigma_n$. We show that these inclusions are strict and, at the same time, all the maps σ_k are surjective. Suppose $\operatorname{Ker}\sigma_k = \operatorname{Ker}\sigma_{k+1}$ for some k. Let π denote the canonical projection of B^{k+1} onto B^k : $\pi(\dot{\sum}_{i=1}^{k+1} x_i) = \dot{\sum}_{i=1}^{k} x_i$. Then obviously $\pi\sigma_{k+1} = \sigma_k$. Since $\operatorname{Ker}\sigma_{k+1} = \operatorname{Ker}\sigma_k$, the restriction of π to $\operatorname{Im}\sigma_{k+1}$ is a monomorphism. Therefore, the homomorphism $\rho : \operatorname{Im}\sigma_k \to \operatorname{Im}\sigma_{k+1}$, the inverse of $\pi|\operatorname{Im}\sigma_{k+1}$, exists. Clearly, $\sigma_{k+1} = \rho\sigma_k$, i.e., $\dot{\sum}_{i=1}^{k+1} Ae_i = \rho(\dot{\sum}_{i=1}^{k} Ae_i)$ for all $A \in A$. Consequently, $Ae_{k+1} = \theta(\dot{\sum}_{i=1}^{k} Ae_i)$ for all $A \in A$, where θ denotes the composition of ρ and the projection from B^{k+1} onto the $(k+1)$st copy of B. To prove the surjectivity of the homomorphism σ_k we proceed by induction (σ_1 is surjective because $Ay = B$; see the first line of the proof). Suppose that σ_k is surjective. Then θ is defined on the whole space B^k and maps it into B. Composing θ with the k canonical imbeddings $B \to B^k$ we obtain operators $\theta_i : B \to B$, $i = 1,\dots,k$, such that $\theta(\dot{\sum}_{i=1}^{k} x_i) = \sum_{i=1}^{k} \theta_i x_i$ for every choice of vectors $x_1,\dots,x_k \in B$. Then $Ae_{k+1} =$ $= \sum_{i=1}^{k} {}_iAe_i$ for all $A \in A$. Applying to both sides of this identity an arbitrary operator $B \in A$, we get $BAe_{k+1} = \sum_{i=1}^{k} B\theta_i Ae_i$. On the other hand, we have $BAe_{k+1} = \sum_{i=1}^{k} \theta_i BAe_i$, since $BA \in A$. Thus, $\sum_{i=1}^{k} B\theta_i Ae_i = \sum_{i=1}^{k} \theta_i BAe_i$, which in view of the surjectivity of σ_k yields

$$\sum_{i=1}^{k} B\theta_i x_i = \sum_{i=1}^{k} \theta_i Bx_i$$

for every choice of vectors $x_1,\dots,x_k \in B$. Consequently, $B\theta_i = \theta_i B$ for $i = 1,\dots,k$ and all $B \in A$. By Schur's Lemma, θ_i are scalar operators : $\theta_i = \lambda_i E$, $i = 1,\dots,k$. It follows that $\theta(\dot{\sum}_{i=1}^{k} x_i) = \sum_{i=1}^{k} \lambda_i x_i$ and so $Ae_{k+1} = \sum_{i=1}^{k} \lambda_i Ae_i$ for all

$A \in \mathcal{A}$. We see that the one-dimensional subspace spanned by the vector $e_{k+1} - \sum_{i=1}^{k} \,_i e_i$ is annihilated by all operators $A \in \mathcal{A}$, and hence is $\mathcal{A}$-invariant, which contradicts the irreducibility of algebra $\mathcal{A}$. We thus showed that if σ_k is surjective, then $\mathrm{Ker}\,\sigma_k \neq \mathrm{Ker}\,\sigma_{k+1}$, i.e., the inclusion $\mathrm{Ker}\,\sigma_k \supset \mathrm{Ker}\,\sigma_{k+1}$ is strict. This means that there is an operator $A \in \mathcal{A}$ such that $Ae_1 = \ldots = Ae_k = 0$, but $Ae_{k+1} \neq 0$. The set $J = \mathrm{Ker}\,\sigma_k$ is a left ideal in $\mathcal{A}$. Therefore, the orbit Je_{k+1} is a nonnull invariant subspace, as we already observed. Since $\mathcal{A}$ is irreducible, $Je_{k+1} = \mathcal{B}$. But this implies that the map σ_{k+1} is surjective. In fact, for every collection $x_1,\ldots,x_k,x_{k+1} \in \mathcal{B}$ we can find (thanks to the surjectivity of σ_k) an $A \in \mathcal{A}$ such that $Ae_1 = x_1$, $\ldots, Ae_k = x_k$, and also (by the preceding argument) a $B \in \mathcal{A}$ such that $Be_1 = \ldots = Be_k = 0$ and $Be_{k+1} = x_{k+1}$. Setting $C = A + B$, we have an operator $C \in \mathcal{A}$ such that $Ce_1 = x_1,\ldots,Ce_k = x_k$, $Ce_{k+1} = x_{k+1}$. This completes the induction. We thus conclude that the map σ_n is surjective, which means that algebra $\mathcal{A}$ contains all the linear operators, as claimed.

□

Burnside's Theorem lies at the foundations of the theory of finite dimensional representations (see Chapter 3).

3°. The most important invariant subspaces of an operator T are connected with its spectrum. Such are, in the first place, the eigensubspaces of T. In addition, with each eigenvalue λ of T one associates the increasing chain of *root subspaces* $W_k = \mathrm{Ker}(T - \lambda E)^k$, $k = 1,2,3,\ldots$, which are obviously T-invariant.

<u>Exercise.</u> *If* $W_{m+1} = W_m$ *for some* m, *then* $W_k = W_m$ *for all* $k \geq m$.

In connection with this one defines the *order* or *rank* of the eigenvalue λ as the largest $m \geq 1$ for which $W_m \neq W_{m-1}$ ($W_0 = 0$) if such an m exists, and as ∞ otherwise. If λ has finite order m, W_m is naturally termed the *maximal* root subspace. In a finite dimensional space $\mathcal{B}$ every eigenvalue has finite order $m \leq \dim \mathcal{B}$. In an infinite dimensional space eigenvalues of infinite order may arise.

THEOREM. *Every eigenvalue* $\lambda \neq 0$ *of a compact operator* T *has finite order and the corresponding maximal root subspace is finite dimensional.*

PROOF. Suppose $W_k \neq W_{k-1}$ for all k. Chose a unit-norm vector $x_k \in W_k$ such that $\mathrm{dist}(x_k, W_{k-1}) > \frac{1}{2}$. Since $Tx_k - \lambda x_k \in W_{k-1}$, $\mathrm{dist}(Tx_k, W_{k-1}) > \frac{1}{2}|\lambda|$. But $x_i \in W_i \subset W_{k-1}$ for all $i < k$, whence $Tx_i \in W_{k-1}$. Consequently, $\mathrm{dist}(Tx_k, Tx_i) > \frac{1}{2}|\lambda|$, which contradicts the compactness of T. Thus, the order m of the eigenvalue λ is finite.

As we already know, $d_1 = \dim W_1 < \infty$. But $(T - \lambda E)^j W_k \subset W_{k-j}$ $(j \leqslant k)$, so denoting $d_j = \dim W_j$ we have $d_k \leqslant d_j + d_{j-k}$. Therefore, $d_m \leqslant md_1 < \infty$.

□

Now let T be an arbitrary operator in B. A subspace $L \subset B$, $L = 0$, is called a *spectral subspace* (and a *spectral maximal subspace* by other authors : transl. note) of T if the following conditions are satisfied :

1) L is invariant under T ;
2) $\mathrm{spec}(T|L) \subset \mathrm{spec}\, T$;
3) if $M \neq 0$ is a T-invariant subspace with the property that $\mathrm{spec}(T|M) \subset \mathrm{spec}(T|L)$, then $M \subset L$.

We call a compact set $Q \subset \mathrm{spec}\, L$ *spectral* if there is a spectral subspace L such that $\mathrm{spec}(T|L) = Q$. The simplest example is of course $Q = \mathrm{spec}\, T$, the spectral compact set corresponding to the spectral subspace $L = B$. It may happen that there are no other spectral compact sets.

<u>Example.</u> Consider the Banach space A of all scalar functions analytic in the disk $|\lambda| < 1$ and continuous in the closed disk $|\lambda| \leqslant 1$, endowed with the norm $\|\phi\| = \max_{|\lambda|=1} |\phi(\lambda)|$. Let Λ denote the operator of multiplication by λ acting in A. The spectrum of Λ coincides with the closed disk $\mathbb{D} = \{\lambda \mid |\lambda| \leqslant 1\}$.

Let $Q \subset \mathbb{D}$ be a nontrivial spectral compact set and let L be the corresponding spectral subspace. Let $\Delta = \mathbb{D} \setminus Q$. If $\mu \in \Delta$, then for every $\phi \in L$ the function $(\lambda - \mu)^{-1}\phi(\lambda)$ must belong to A, and then $\phi(\mu) = 0$. Thus, $\phi|\Delta = 0$ for all $\phi \in L$. But then $\phi = 0$ by the uniqueness theorem for analytic functions, i.e., $L = 0$, contrary to the definition of a spectral subspace.

THEOREM. *Let* T *be a compact operator. Then every point* $\lambda \in \text{spec}(T)$, $\lambda \neq 0$, *is a spectral compact set. The corresponding spectral subspace is equal to the maximal root subspace* W_m.

PROOF. Since W_m is T-invariant and $(T - \lambda E)^m | W_m = 0$, $\text{spec}(T|W_m) = \{\lambda\}$. It remains to verify that if $M \neq 0$ is a T-invariant subspace such that $\text{spec}(T|M) = \{\lambda\}$, then $M \subset W_m$. But $T|M$ is compact and $0 \notin \text{spec}(T|M)$. Consequently, $\ell = \dim M < \infty$, and then, by the Cayley-Hamilton theorem, $(T - \lambda E)^{\ell} M = 0$. Therefore, $M \subset W_{\ell}$, and since W_m is the maximal root subspace corresponding to λ, $M \subset W_m$.

□

Exercise. *Suppose* T *is a compact operator. Then every finite subset* $F \subset \text{spec}(T)$ *such that* $0 \notin F$ *is a spectral compact set and the corresponding spectral subspace is equal to the sum of the maximal root subspaces corresponding to the points* $\lambda \in F$.

Using the *operational (functional) calculus* (an exposition of which can be found in [41]), one can show that *if* $T \in L(B)$ *and the compact set* Q *is open in* spec T, *then* Q *is a spectral compact set (here* spec T *must be disconnected).*

Let T *be a self-adjoint operator in Hilbert space. Then every compact set* $Q \subset \text{spec}\, T$ *which has a nonempty interior (as a subset of the topological space* spec T*) is spectral.* This is an easy consequence of the general spectral theory of self-adjoint operators (for the study of which we recommend the monographs [1] and [41]). An analogous statement holds true for operators which are spectral in the sense of N. Dunford ([13] is devoted to the theory of such operators).

The operator $T \in L(B)$ is called an *operator with separable spectrum* if the family of its spectral compact sets is a basis for the topology of spec T, meaning that every set open in spec T is a union of interiors of spectral compact sets. For this to happen it suffices that every compact set which is the closure of its interior be a spectral compact set. [Here we followed the original terminology of the author. The closely related notion used in the western literature is that of a *decomposable operator* ; see, for example, [8] and [13] ; transl. note.]

THEOREM (Lyubich-Matsaev, 1960). *Suppose that the spectrum of the operator* T *lies on a smooth curve* C. *Suppose further that for every point* $\mu \in \text{spec}\, T \subset C$ *the inequality* $\|R_\mu\| \leqslant M_\mu(\text{dist}(\lambda, C))$ *holds in a neighborhood of* μ, *where* $M_\mu(\delta)$ *is a decreasing function of* $\delta > 0$ *obeying the condition*

$$\int_0^{\varepsilon} \ln \ln M_\mu(\delta)\, d\delta < \infty$$

for any sufficiently small $\varepsilon > 0$. *Then* T *is an operator with separable spectrum.*

[The integral condition on the majorant $M_\mu(\delta)$ goes back to a theorem of N. Levinson (1940), and is known as the *Levinson condition.*]

The proof of this theorem relies on a difficult analytic technique that goes beyond the scope of the present book. The reader may consult the paper of Yu. I. Lyubich and V. I. Matsaev (1962) for a detailed exposition of the proof. For operators with real (or unimodular, i.e., contained in the unit circle) spectrum the problem of the separability of the spectrum can be formulated and solved in terms of representation theory. This leads to an alternate proof discussed in Chapter 5.

Under specific conditions, a duality of one or other kind related to the general operator-theoretic duality between images and kernels, holds in the theory of spectral subspaces. This aspect is considered in a remarkable work of E. Bishop (1959) and a related paper of V. I. Lomonosov, Yu. I. Lyubich, and V. I. Matsaev (1974). Extensive and far reaching investigations in the theory of spectral subspaces were undertaken over the last 20 years by C. Foiaş and his school (this direction is treated in [8]).

5. COMMUTATIVE BANACH ALGEBRAS

1°. The theory of commutative Banach algebras was founded by I. M. Gelfand in the end of the thirties. As it turned out straight away, in not only has intrinsic depth, but is also fruitful in applications, in particular, in analysis (I. M. Gelfand, 1939) and representation theory (I. M. Gelfand and D. A. Raikov, 1940 ; M. G. Krein, 1949).

All Banach algebras considered in this section are tacitly assumed to be commutative. Concerning the norm we assume that $\|xy\| \leq \|x\| \|y\|$ and $\|e\| = 1$, where e is the unit of the algebra ; this can be always achieved by replacing the given norm with an equivalent norm.

Let A be a Banach algebra. The *spectrum* spec x of the element $x \in A$ is, by definition, the spectrum of the operator $R_x \in L(A)$ given by $R_x y = xy$. The *domain of regularity* (or the *resolvent set*) reg x of x is defined in similar manner. By the result obtained previously for operators, *the domain of regularity of any element* x *is open, while its spectrum is nonempty and compact.* The *spectral radius* $\rho(x)$ of $x \in A$ is, by definition, the spectral radius of the operator R_x. Since $\|R_x\| = \|x\|$ and $R_{x^k} = R_x^k$, Gelfand's formula carries over to x :

$$\rho(x) = \lim_{k\to\infty} \|x^k\|^{1/k} \quad (= \inf_k \|x^k\|^{1/k}) .$$

An element x such that $\rho(x) = 0$ is called a *quasi-nilpotent*. In particular, such is every nilpotent (i.e., every x such that $x^k = 0$ for some k).

Exercise. *$\lambda \in$ reg x if and only if the element $x - \lambda e$ is invertible.*

Let us examine closer the set (group) of all invertible elements.

LEMMA. *Every element in the ball* $\|x - e\| < 1$ *is invertible.*

PROOF. If x is not invertible, then $0 \in \operatorname{spec} x$, and then $-1 \in \operatorname{spec}(x - e)$. Consequently, $\rho(x - e) \geqslant 1$, whence $\|x - e\| \geqslant 1$.

$\square$

COROLLARY. *The set of all invertible elements is open.*

PROOF. If x is invertible and $\|y\| < \|x^{-1}\|^{-1}$, then $x + y$ is invertible, since $x + y = x(e + x^{-1}y)$ and $\|x^{-1}y\| \leqslant \|x^{-1}\|\|y\| < 1$.

$\square$

A key problem is to characterize those Banach algebras in which every nonzero element is invertible (i.e., the *Banach fields*).

THE GELFAND-MAZUR THEOREM. *A Banach field consists only of scalar elements (i.e., elements* λe *with* $\lambda \in \mathbb{C}$).

PROOF. For each x there exists a point $\lambda \in \operatorname{spec} x$. Then $x - \lambda e$ is not invertible, and so $x - \lambda e = 0$, i.e., $x = \lambda e$, as claimed.

$\square$

Thus, *every Banach field can be naturally identified with the complex field* $\mathbb{C}$. This fact is the basis of the constructions that follow.

2°. A basic object in Gelfand's theory is the so-called maximal ideal space of a Banach algebra. The notion of a *maximal ideal* itself is purely algebraic : it is a proper (i.e., different from the full algebra) ideal which is not contained in any proper ideal. [For Banach algebras the terms "ideal" and "subalgebra" have the usual algebraic meaning, i.e., closedness is not a requirement. Our convention is that every subalgebra necessarily contains the unit e, whereas an ideal different from the full algebra does not contain e.] The quotient of the given algebra by any maximal ideal is a field. Conversely, every ideal for which the corresponding quotient algebra is a field is maximal. The existence of maximal ideals is readily established using Zorn's Lemma. The same tool permits us to conclude that every proper ideal is contained in a maximal ideal. This shows that the union of all maximal ideals of an algebra is equal to the set of its noninvertible elements. All these algebraic facts are independent of the Banach structure. The latter puts its mark only of the geometric nature of maximal ideals, but this mark is rather noticeable.

THEOREM. *Let* A *be a Banach algebra. Then :*

1) every maximal ideal of A *is closed ;*

2) every maximal ideal of A *is the kernel of a unique multiplicative linear functional (i.e., of a continuous homomorphism of algebra* A *into the field* $\mathbb{C}$*);*

3) conversely, the kernel of every multiplicative linear functional is a maximal ideal.

[From now on we shall omit the adjective "linear" and say simply "multiplicative functional". Incidentally, the proof of the theorem shows that *every homomorphism* $A \to \mathbb{C}$ *is continuous*.]

PROOF. 1) It follows from the continuity of the addition and multiplication operations in A that the closure of any ideal is again an ideal. If M is a maximal ideal, then the only case where the ideal $\overline{M} \supset M$ is not equal to M is $\overline{M} = A$. But the latter is impossible since M is contained in the complement of the nonempty set of all invertible elements.

2) Consider the quotient algebra A/M. It is a field, and in fact a Banach field with respect to the standard norm on the quotient space. By the Gelfand-Mazur theorem, $A/M = \{\lambda[e]\}_{\lambda\in\mathbb{C}}$ (for $x \in A$ we let $[x]$ denote the class of x mod M). The map $\lambda[e] \to \lambda$ is a multiplicative functional on A/M, and its composition with the canonical morphism $A \to A/M$ yields a multiplicative functional f_M on M. Obviously, $M = \operatorname{Ker} f_M$. If now g is any linear functional on A with $\operatorname{Ker} g = M$, then $g = \alpha f_M$; g is

multiplicative only for $\alpha = 1$.

[Note. The class of Banach algebras is a category in which the *morphisms* are, by convention, the continuous homomorphisms. Since the existence of a unit is for us part of the definition of a Banach algebra, morphisms must preserve the unit.]

3) Let f be a multiplicative functional on A. Then $M = \text{Ker } f$ is an ideal, as is the kernel of any morphism. Moreover, M is maximal : in fact, $M \neq A$, and if the ideal $J \supset M$ is different from M, there is an $x_0 \in J$ such that $f(x_0) \neq 0$, and then $x - [f(x_0)]^{-1} f(x) x_0 \in M$ for all $x \in A$, whence $A \subset J$, i.e., $J = A$.

□

Thus, $f \to \text{Ker } f$ is a bijection of the set of all multiplicative functionals onto the set of all maximal ideals, which permits us to identify maximal ideals with the corresponding multiplicative functionals.

LEMMA. *The norm of any multiplicative functional* f *is equal to one.*

PROOF. We have $|f(x)| = |f(x^k)|^{1/k} \leqslant \{\|f\| \|x^k\|\}^{1/k}$, $k = 1, 2, 3, \ldots$, whence, by Gelfand's formula, $|f(x)| \leqslant \rho(x)$. This in turn gives $|f(x)| \leqslant \|x\|$. On the other hand, $f(e) = 1$. Thus, $\|f\| = 1$, as claimed.

□

The inequality $|f(x)| \leqslant \rho(x)$ obtained in the last proof is in fact more important than the final result. A particular consequence of it is that if x is a quasi-nilpotent, then $f(x) = 0$ for all multiplicative functionals f, i.e., x belongs to the *radical* of algebra A, defined as the intersection of all maximal ideals of A. Actually, we have the following result.

THEOREM. *The radical is equal to the set of all quasi-nilpotents.*

PROOF. In view of the preceding remark, all we have to check is that $\rho(x) = 0$ whenever x belongs to the radical. Let

$\lambda \neq 0$. Then $x - \lambda e$ does not belong to any maximal ideal, since x belongs to all of them, while e belongs to none. Therefore, $x - \lambda e$ is invertible, i.e., $\lambda \in \operatorname{reg} x$. Consequently, $\operatorname{spec} x = \{0\}$, as asserted.

□

A Banach algebra is called *semisimple* if its radical is equal to zero. *The quotient of any Banach algebra by its radical is semisimple.*

COROLLARY 1. *Every closed subalgebra of a semisimple Banach algebra is semisimple.*

□

COROLLARY 2. *Let* A *be a Banach algebra such that* $\rho(x) = \|x\|$ *for all* $x \in A$. *Then* A *is semisimple.*

□

Consider now the set $M(A)$ of all maximal ideals of the Banach algebra A. Identifying maximal ideals with the corresponding multiplicative functionals, we realize $M(A)$ as a subset of the unit ball in the conjugate space A^*. The unit ball $\|f\| \leqslant 1$ in A^* is compact in the w^*-topology. Moreover, $M(A)$ is closed in this topology, since it is given by the system of equations $f(x + y) = f(x) + f(y)$, $f(xy) = f(x)f(y)$ $(\forall\ x,y \in A)$, $f(\alpha e) = \alpha f(e)$ $(\forall\ \alpha \in \mathbb{C})$. We thus have the following result.

THEOREM. *The set* $M(A)$ *of all maximal ideals of the Banach algebra* A, *regarded as a topological subspace of the conjugate space* A^* *endowed with the* w^*-*topology, is compact.*

□

The compactum $M(A)$ is called the *maximal ideal space* of the Banach algebra A. The mapping $A \to M(A)$ may be treated as a contravariant functor from the category of Banach algebras into the category of compact topological spaces. In fact, if A_1, A_2 are Banach algebras and $h : A_1 \to A_2$ is a morphism, then the induced map $h^* : M(A_2) \to M(A_1)$, given by $h^*f = f \circ h$, is continuous. Also, $\mathrm{id}^* = \mathrm{id}$ and $(h_2h_1)^* = h_1^*\, h_2^*$ for arbitrary morphisms $h_1 : A_1 \to A_2$ and $h_2 : A_2 \to A_3$.

The usefulness of this remark is evident. An immediate consequence is that *to isomorphic Banach algebras there correspond homeomorphic maximal ideal spaces.*

The next example is of fundamental importance.

Example. Let S be a compact topological space. The Banach space $C(S)$ of continuous functions on S is a Banach algebra under pointwise multiplication (and in $C(S)$, $\|\phi\psi\| \leq \|\phi\| \|\psi\|$, $\|1\| = 1$). Every point $s \in S$ defines a multiplicative functional δ_s on $C(S)$ by the rule $\delta_s(\phi) = \phi(s)$. The correponding maximal ideal is $M_s = \{\phi \mid \phi(s) = 0\}$. This yields the *canonical mapping* $\delta : S \to M$ $(= M(C(S)))$.

THEOREM. *δ is a homeomorphism.*

□

The proof of this theorem is left to the reader.

Henceforth the maximal ideal space of the algebra $C(S)$ will be identified with S through the canonical homeomorphism δ. If S_1, S_2 are compact spaces and $h : C(S_1) \to C(S_2)$ is a Banach algebra morphism, then the induced continuous map $h^* : S_2 \to S_1$ is such that $h(\phi) = \phi \circ h^*$. Moreover, h^* is a homeomorphism whenever h is an isomorphism.

In the Banach algebra $C(S)$ the spectrum of any element ϕ coincides with the range $\{\phi(s) \mid s \in S\}$ of the function ϕ; indeed, $\phi - \lambda$ is invertible if and only if this function does not vanish on S. Accordingly, *the spectral radius of ϕ is* $\rho(\phi) = \max_s |\phi(s)| = \|\phi\|$. Therefore, *algebra $C(S)$ is semisimple.*

Now let A be an arbitrary Banach algebra. We assign to each element $x \in A$ the function $\tilde{x}$ on $M(A)$ defined by the formula $\tilde{x}(f) = f(x)$ (in other words, we restrict to $M(A) \subset A^*$ the image of x under the canonical mapping $A \to A^{**}$).

THEOREM. *The mapping $A \to C(M(A))$, $x \to \tilde{x}$, is a morphism of Banach algebras. Its kernel is equal to the radical of A. Its image is a subalgebra of $C(M(A))$ that separates the points of the*

the compactum $M(A)$.

PROOF. The required algebraic properties are obvious. The continuity of the mapping $x \to \tilde{x}$ follows from the inequality $\|\tilde{x}\| = \max_f |f(x)| \leq \|x\|$. The equality $\tilde{x} = 0$ means that $f(x) = 0$ for all $f \in M(A)$, i.e., that x belongs to the radical of A. The image $\tilde{A} = \{\tilde{x} \mid x \in A\} \subset C(M(A))$ is a subalgebra, as is the image of any morphism of algebras. Finally, if $f_1, f_2 \in M(A)$, $f_1 \neq f_2$, then $\operatorname{Ker} f_1 \not\subset \operatorname{Ker} f_2$, and hence there is an $x \in A$ such that $f_1(x) = 0$, $f_2(x) \neq 0$, i.e., $\tilde{x}(f_1) = 0$, $\tilde{x}(f_2) \neq 0$; the function $\tilde{x}$ separates the points f_1, f_2.

□

The morphism described in the theorem and its image are called the *Gelfand representation* (or *transformation*) and, respectively, the *Gelfand image of algebra* A.

COROLLARY. *The Gelfand representation of* A *is injective if and only if* A *is semisimple.*

□

In this case A can be identified with its Gelfand image and accordingly regarded as a subalgebra of the algebra of all continuous functions on a compact space.

THEOREM. *The spectrum of the element* $x \in A$ *coincides with the range of its Gelfand representative* $\tilde{x}$, *i.e., with the spectrum of the function* $\tilde{x}$ *in* $C(M(A))$.

PROOF. $x \in A$ is invertible if and only if $x \notin M$ for every maximal ideal M, i.e., if and only if $f(x) \neq 0$ for every multiplicative functional f. Now replace x by $x - \lambda e$ and notice that $(x - \lambda e)^{\sim} = \tilde{x} - \lambda$.

□

COROLLARY. $\|\tilde{x}\| = \rho(x)$ *for all* $x \in A$.

□

<u>Exercise.</u> *The Gelfand representation is an isometry of the Banach algebra* A *onto its Gelfand image if and only if* $\|x^2\| =$

$= ||x||^2$ *for all* $x \in A$.

3°. We now address the following important question. Let S be a compact space. What intrinsic properties should a subalgebra A of the Banach algebra $C(S)$ enjoy in order that it be dense in $C(S)$? The adequate form of the answer to this question encompasses the classical theorem of Weierstrass on uniform approximation of functions continuous on a segment by polynomials (in which the role of A is played by the subalgebra of all polynomials). The general problem was solved by M. Stone in 1937.

THE STONE-WEIERSTRASS THEOREM. *Let* $A \subset C(S)$ *be a subalgebra. Suppose that :*

1) A is symmetric, i.e., if $\phi \in A$, then the complex-conjugate function $\overline{\phi} \in A$;

2) the functions in A separate the points of the compactum S.

Then A is dense in $C(S)$.

We remark that property 2) is necessary because $C(S)$ enjoys it.

For the proof (and also independently) it is useful to state and prove

THE REAL VARIANT OF THE STONE-WEIERSTRASS THEOREM. *Let $C_{\mathbb{R}}(S)$ denote the Banach algebra of all continuous real-valued functions on the compact space S, and let A be a subalgebra of $C_{\mathbb{R}}(S)$. If the functions in A separate the points of S, then A is dense in $C_{\mathbb{R}}(S)$.*

PROOF (L. de Branges, 1959). Suppose A is not dense. Let N denote the set of all real (generally speaking, sign-alternating) measures ν on S which annihilate A and satisfy the condition $||\nu|| \leqslant 1$. This is a convex, centrally-symmetric compact subset of the space of all measures on S endowed with the w^*-topology. Since A is not dense, $N \neq \{0\}$. By the Krein-Milman Theorem, N has an extreme point $\sigma \equiv ds$. Obviously, $||\sigma|| = 1$ and, by construction, $\int \phi\, ds = 0$ for all $\phi \in A$; in particular, $\int ds = 0$. It follows from this that the support of σ contains at least two distinct points s_1, s_2. By hypothesis, there is a function $\psi \in A$ which separates s_1 and s_2 : $\psi(s_1) \neq \psi(s_2)$. We may assume, with no loss of generality, that $0 < \psi(s) < 1$ for all $s \in S$. In fact, if $\alpha = \min \psi$ and $\beta = \max \psi$, then the function $(\beta - \alpha + 2)^{-1}(\psi(s) - \alpha + 1)$ satisfies the required inequality, belongs to A, and separates the points s_1 and s_2. Since A is a subalgebra, $\phi\psi \in A$ for all $\phi \in A$, and so $\int \phi\psi\, ds = 0$. Thus, the measure $dt = \psi ds$ annihilates A, and hence the same is true for the measure $dr = ds - dt = (1 - \psi)ds$. Obviously, dt and dr are different from zero, and can therefore be normalized. Let $\hat{dt} = dt(\int |dt|)^{-1}$ and $\hat{dr} = dr(\int |dr|)^{-1}$. Then $ds = dt + dr = p\,\hat{dt} + q\,\hat{dr}$, where $p > 0$, $q > 0$, and $p + q = \int |dt| + \int |dr| = \int \psi |ds| + \int (1 - \psi)|ds| = \int |ds| = \|\sigma\| = 1$. But ds is an extreme point for N, and $\hat{dt}$, $\hat{dr} \in N$. It follows that $\hat{dt} = a\, ds$, where $a = \text{const}$, i.e., $\psi ds = b\, ds$, where $b = \text{const}$. We arrived at a contradiction : ψ is constant on the support of σ, whereas $\psi(s_1) \neq \psi(s_2)$.

□

Now to prove the Stone-Weierstrass Theorem it suffices to consider the set $\operatorname{Re} A = \{\theta \mid \theta = \operatorname{Re}\phi,\ \phi \in A\}$. If $\phi \in A$, then by hypothesis $\overline{\phi} \in A$, and so $\operatorname{Re}\phi \in A$, too. Therefore, $\operatorname{Re} A \subset A$, and $\operatorname{Re} A$ is a subalgebra of the real algebra $C_{\mathbb{R}}(S)$ that separates the points of the compactum S. By the Real Stone-Weierstrass Theorem, $\operatorname{Re} A$ is dense in $C_{\mathbb{R}}(S)$, which in turn implies that A is dense in $C(S)$.

□

A Banach algebra is said to be *symmetric* if its Gelfand image is symmetric.

COROLLARY. *Let* A *be a symmetric Banach algebra. Then the Gelfand image of* A *is dense in* $C(M(A))$.

□

4°. The Banach algebra A is called *regular* if for every compact set $Q \subset M(A)$ and every point $f_0 \in M(A)$ there is an $x \in A$ such that $\tilde{x}|Q = 0$ and $\tilde{x}(f_0) \neq 0$. The merit for studying this important class of algebras goes to G. E. Shilov (see his monograph [47]). The simplest example of regular Banach algebra is, by Urysohn's Lemma, $C(S)$, where S is a compact space. An analogue of Urysohn's Lemma holds true in every regular algebra.

THEOREM. *Let* A *be a regular Banach algebra and let* Q, K *be compact subsets of* $M(A)$. *Then there exists an* $x \in A$ *such that* $\tilde{x}|Q = 0$ *and* $\tilde{x}|K = 1$.

For the proof we need the following

LEMMA. *Let* A *be a regular algebra,* K *a compact subset of* $M(A)$, *and* $I(K)$ *the closed ideal defined by the condition* $\tilde{x}|K = 0$. *Then the maximal ideal space of the Banach algebra* $A/I(K)$ *is canonically homeomorphic to* K.

PROOF OF THE LEMMA. Consider the canonical homomorphism $j : A \to A/I(K)$. It induces the continuous mapping $j^* : M(A/I(K)) \to M(A)$ which is injective thanks to the surjectivity of j. Im j^* is equal to the set of those maximal ideal of A which contain $I(K)$. Since A is regular, this set coincides with K. Consequently, j^* is a continuous bijection of the compact spaces $M(A/I(K))$ and K, and hence a homeomorphism.

□

PROOF OF THE THEOREM. Consider the ideals $I(Q)$ and $I(K)$. The image J of $I(Q)$ under the canonical homomorphism $j : A \to A/I(K)$ is an ideal thanks to the surjectivity of j. By the lemma, the maximal ideals of $A/I(K)$ can be identified with the points of K. Since A is regular, given any $f_0 \in K$ there is an $x \in I(Q)$ such that $\tilde{x}(f_0) \neq 0$, i.e., $jx \in J$ does not belong to the maximal ideal of $A/I(K)$ corresponding to f_0. We see that J is not contained in any maximal ideal of $A/I(K)$. Consequently, $J = A/I(K)$, and then the unit belongs to J. Any preimage of the unit is an element $x \in I(Q)$ with the property that $\tilde{x}|K = 1$;

at the same time, $\tilde{x}|Q = 0$.

□

Remark. *If under the hypotheses of the theorem* A *is symmetric, then* x *can be choosen so that* $0 \leqslant \tilde{x}(f) \leqslant 1$ *for all* f. In fact, let x be the element provided by the theorem and let x^* be any element whose Gelfand transform is the complex-conjugate of $\tilde{x}$. Set $v = xx^*$. Then $v = |x|^2 \geqslant 0$, $\tilde{v}|Q = 0$, and $\tilde{v}|K = 1$. Let $\mu = \max_f |\tilde{v}(f)|$. Pick a polynomial $p(\tau)$ satisfying $p(0) = 0$, $p(1) = 1$, and $|p(\tau)| \leqslant 1$ for all $0 \leqslant \tau \leqslant \mu$. Then $z = p^2(v)$ enjoys all the needed properties.

Exercise. *Let* A *be a regular symmetric Banach algebra. Then for every point* $f_0 \in M(A)$ *there is an* $x \in A$ *such that* $\tilde{x}(f_0) = 1$ *and* $|\tilde{x}(f)| < 1$ *for all* $f \neq f_0$ (the terminology is that f_0 is a *peak point* for A).

An important property of regular Banach algebras is given by the following theorem of G. E. Shilov (1940) on the extension of multiplicative functionals.

THEOREM. *Let* A *be a regular Banach algebra which is also a closed subalgebra of a Banach algebra* A°. *Then every functional* $f_0 \in M(A)$ *extends to a functional* $f_0^\circ \in M(A^\circ)$.

[In point of fact, even if A is not regular, the extension property remains valid for the functionals belonging to the so-called *Shilov boundary*.]

PROOF. Consider in A the maximal ideal $M_0 = \operatorname{Ker} f_0$ and the smallest ideal J in A° that contains M_0. If $J \neq A^\circ$, then $J \subset M_0^\circ$, where M_0° is a maximal ideal in A°. Let f_0° be a multiplicative functional on A° such that $M_0^\circ = \operatorname{Ker} f_0^\circ$. Now let $x \in A$. Then $x = f_0(x)e + y$, where $y \in M_0 \subset M_0^\circ$. Hence, $f_0^\circ(x) = f_0(x)$, i.e., $f_0^\circ A = f_0$. Now suppose that $J = A^\circ$. We claim that this leads to a contradiction. In fact, if $J = A^\circ$, then $e \in J$, i.e., $e = \sum_{i=1}^n x_i y_i$, with $x_i \in M_0$ and $y_i \in A^\circ$.

We may assume, with no loss of generality, that $\|x_i\| = 1$, $1 \leq i \leq n$. Set $\mu = \max_i \|y_i\|$ and let N denote the neighborhood of f_0 in $M(A)$ defined by the inequalities $|\tilde{x}_i(f)| < (2n\mu)^{-1}$. Since A is regular, there is a $z \in A$ such that $\tilde{z}(f) = 0$ for all $f \in N$ and $\max_f |\tilde{z}(f)| = 1$. Then $|(zx_i)^{\sim}(f)| < (2n\mu)^{-1}$ for all $f \in M(A)$ and $1 \leq i \leq n$. But $z = \sum_{i=1}^n (zx_i)y_i$, and so if $g \in M(A^\circ)$, then $\hat{g} \equiv g|A \in M(A)$ and $\tilde{z}(g) = \sum_{i=1}^n (zx_i)^{\sim}(\hat{g})y_i(g)$, whence $|\tilde{z}(g)| \leq \sum_{i=1}^n (2n\mu)^{-1}\mu = \frac{1}{2}$. Therefore, the spectral radius of z in A does not exceed $1/2$. However, by Gelfand's formula, the spectral radius of an element in a subalgebra is equal to its spectral radius in the full algebra. But, in A, $\rho(z) = 1$, because $\rho(z) = \max_{f \in M(A)} |\tilde{z}(f)|$, which completes the proof of the theorem.

□

For further familiarization with the theory and applications of Banach algebras the reader may consult the fundamental monograph of I. M. Gelfand, D. A. Raikov, and G. E. Shilov [15] and also, among others, the books of M. A. Naimark [35] (where the noncommutative case is treated in sufficient depth ; in connection with this see also J. Diximier's book [12]), L. Loomis [32], and N. Bourbaki [5].

CHAPTER 2

TOPOLOGICAL GROUPS AND SEMIGROUPS

1. TOPOLOGICAL GROUPS

1°. A *topological group* is a group endowed with a Hausdorff topology relative to which the operations of multiplication and inversion are continuous (the latter being therefore a homeomorphism) ; here the Cartesian product of the group with itself is endowed with the product topology. [The Hausdorff requirement is not included by all authors in the definition of a topological group. It can be in fact relaxed, without restricting the class of groups, to axiom T_0 : for any two distinct points there is a neighborhood of one of them that does not contain the other.]

Every group equipped with the discrete topology is a topological group and in this quality it is called a *discrete group*. A topological group is discrete if and only if its identity element is isolated. The standard example is the additive group $\mathbb{Z}$ of all integers.

We give several other examples of topological groups (in Examples 1-4 the topology is the standard one).

Example 1. The additive group of the field $\mathbb{C}$ or $\mathbb{R}$ of complex or real numbers, respectively.

Example 2. The multiplicative groups $\mathbb{C}' = \mathbb{C} \setminus \{0\}$ and $\mathbb{R}' = \mathbb{R} \setminus \{0\}$.

Example 3. The unit circle $\mathbb{T}$.

Example 4. The additive group of a (complex or real) Banach space.

Example 5. The group of automorphisms of a Banach space, endowed with the uniform topology (in particular, $GL(n,\mathbb{C})$ and $GL(n,\mathbb{R})$).

Example 6. The group of all invertible elements of a Banach algebra A (with the topology induced from A).

One of the important constructions in group theory, the direct product, admits a canonical topology. In this way $\mathbb{R}$ yields the additive group $\mathbb{R}^n$, and the unit circle $\mathbb{T}$ yields the m-dimensional torus $\mathbb{T}^m$. Then one can build, say, $\mathbb{R}^n \times \mathbb{T}^m \times \mathbb{Z}^\ell \times F$, where F is any finite Abelian group. This already provides a wide class of Abelian topological groups (known as the *elementary groups* ; in particular $\mathbb{Z}^\ell \times F$ is, by the classical structure theorem, the general form of finitely-generated Abelian groups).

From now on by "group" we mean "topological group". Let G be a group. Every element $h \in G$ defines the *left translation by* h, i.e., the mapping $L_h g = h^{-1}g$, and the *right translation by* h, $R_h g = gh$. *Both* L_h *and* R_h *are homeomorphisms.* This implies that *if* D *is an open set in* G *and* $M \subset G$ *is arbitrary, then* MD *and* DM *are open.* In fact, MD, say, can be written as $MD = \cup_{h \in M} L_{h^{-1}} D$.

If $D \subset G$ *is open, then so is* D^{-1}. Consequently, if D is a neighborhood of the identity element e, then so is $\Delta = D \cap D^{-1}$; in addition, Δ is *symmetric*, i.e., if $g \in \Delta$, then $g^{-1} \in \Delta$. Therefore, *every neighborhood of* e *contains a symmetric such neighborhood.*

Subgroups of topological groups endowed with the induced topology are topological groups. As an example we mention the unitary group $U(H)$ of any Hilbert space H (in particular, $U(n)$, $O(n)$) and, more generally, the isometry group Iso B of any Banach space B.

Exercise 1. *The closure of any subgroup (normal subgroup) is a subgroup (respectively, a normal subgroup).*

Exercise 2. *Every discrete subgroup is closed.*

Exercise 3. *The center of any group is closed.*

For any subgroup Γ of the group G one can consider the space G/Γ of right (for the sake of definiteness) cosets of Γ in G [since various terminologies are used in the literature, we must specify that for us the *right (left) coset of* $g \in G$ *modulo* Γ *is the set* $g\Gamma$ *(respectively,* Γg*)*]. We endow G/Γ with the usual quotient topology, i.e., the strongest topology in which the canonical map $j : G \to G/\Gamma$ is continuous.

LEMMA. *The canonical map* $j : G \to G/\Gamma$ *is open.*

PROOF. The preimage of the set jM under j coincides with $M\Gamma$ for every $M \subset G$. Hence, if M is open, then so is $j^{-1}(j(M))$, and then jM is open in G/Γ by the definition of the quotient topology.

□

The topology on G/Γ is not necessarily Hausdorff.

Example. Let $\mathbb{Q}$ be the subgroup of rational numbers in $\mathbb{R}$. If $D \subset \mathbb{R}$ is open, $D \neq \emptyset$, and for each $\rho \in D$ all the numbers $\rho + \sigma$ with $\sigma \in \mathbb{Q}$ belong to D, then $D = \mathbb{R}$. Therefore, the topology on $\mathbb{R}/\mathbb{Q}$ is anti-discrete.

In contrast with this example we have the following

THEOREM. *If the subgroup* $\Gamma \subset G$ *is closed, then* G/Γ *is Hausdorff.*

PROOF. Let $g_1, g_2 \in G$ have distinct right cosets mod Γ, i.e., $g_1^{-1}g_2 \in \Gamma$. Since the complement of Γ in G is open, it follows from the continuity of the operation of left division that g_1 and g_2 admit neighborhoods N_1 and N_2, respectively, such that $(N_1^{-1}N_2) \cap \Gamma = \emptyset$. But then $[(N_1\Gamma)^{-1}(N_2\Gamma)] \cap \Gamma = \emptyset$, whence $N_1\Gamma \cap N_2\Gamma = \emptyset$. Consider the images $M_1 = jN_1$ and $M_2 = jN_2$ of N_1 and N_2 in G/Γ. They are open and disjoint (as their pre-images are disjoint). Hence, they separate the cosets jg_1 and jg_2.

□

COROLLARY. *Let* Γ *be a normal subgroup of* G. *Then* G/Γ *is a topological group.*

□

Exercise. *Let* $\Gamma \subset G$ *be a subgroup such that* G/Γ *is Hausdorff. Then* Γ *is closed.*

It is an interesting fact that *every open subgroup* Γ *is closed.* In fact, any coset $g\Gamma$, $g \in G$, is open, and then so is the union of the cosets different from Γ, i.e., the complement of Γ. Therefore, *a connected group* G *contains no open subgroups* $\Gamma \neq G$.

If the subgroup Γ *is open, then the coset space* G/Γ *is discrete* : its points are images of open sets (the cosets), and hence are open. Thus, we have the following

THEOREM. *Let* Γ *be an open normal subgroup of* G. *Then the quotient group* G/Γ *is discrete.*

□

Exercise. *Let* $\Gamma \subset G$ *be a subgroup such that* G/Γ *is discrete. Then* Γ *is open.*

Let G be a group. Denote by G_0 the component of the identity element e. It is connected by definition and also, as we know, closed.

THEOREM. *The component* G_0 *of* G *is a closed normal sub-*

group. The quotient group G/G_0 *is totally disconnected.*

PROOF. For every $g \in G$ the left translation homeomorphism L_g maps G_0 into a connected set containing e. Consequently, $g^{-1}G_0 \subset G_0$, and so G_0 is a subgroup. Furthermore, for each $g \in G$, the conjugation homeomorphism $L_g R_g$ also maps G_0 into a connected set containing e, and so $g^{-1}G_0 g \subset G_0$. Therefore, the subgroup G_0 is normal. Since the component of any $g \in G$ is its coset gG_0, it follows that in G/G_0 the components reduce to single points, i.e., G/G_0 is totally disconnected.

□

Now let N be an arbitrary neighborhood of e in G. *The subgroup* $\Gamma(N)$ *generated by* N *is open (and hence closed).* In fact, it can be represented as $\Gamma(N) = \bigcup N^{\varepsilon_1}\dots N^{\varepsilon_k}$, with $\varepsilon_i = \pm 1$, $k = 1,2,3,\dots$. Therefore, *a connected group* Γ *is generated by any neighborhood of the identity. If the neighborhood* N *of* e *is connected, then so is the subgroup* $\Gamma(N)$. In fact, the set $N^{\varepsilon_1}\dots N^{\varepsilon_k}$ is the image of the connected set $N \times \dots \times N$ under the continuous map $(g_1,\dots,g_k) \to g_1^{\varepsilon_1}\dots g_k^{\varepsilon_k}$. Moreover, e $N^{\varepsilon_1}\dots N^{\varepsilon_k}$. Thus, $\Gamma(N)$ is a union of connected sets that have a common point, and hence is connected. In the end we obtain the following

THEOREM. *Suppose the group* G *contains a connected neighborhood* N *of the identity. Then the component* G_0 *of the identity is generated by* N *and is a connected open normal subgroup. The quotient group* G/G_0 *is discrete.*

PROOF. $\Gamma(N) \subset G_0$, and so N is a neighborhood of e in G_0. Since G_0 is connected, it is generated by N, i.e., $G_0 = \Gamma(N)$. Therefore, G_0 is open and, by the preceding theorem, it is also closed and normal. Finally, G/G_0 is discrete because G_0 is open.

□

Exercise 1. *In a connected group every totally disconnected normal subgroup is contained in the center.*

Exercise 2. *The commutator subgroup of a connected group is connected.*

We next give a rather eccentric example of a connected group that will be encountered several times in the sequel.

Example. Consider the group of all measurable subsets of the segment $[0,1]$ (subsets of measure zero are negligible) with the operation $\oplus$ of symmetric difference and the topology induced by the metric $\text{mes}(M \oplus N)$ (where $\text{mes}(\cdot)$ denotes the Lebesgue measure). *Every element of this group has order 2 (it is a 2-group). However, the group is connected (and even path connected);* this follows from the fact that for every element M the continuous curve $M \cap [0,\tau]$, $0 \leqslant \tau \leqslant 1$, connects the empty set (which serves as the identity element of the group) with M.

The most important class of topological groups is that of the *Lie groups*. These are defined as (real or complex) analytic manifolds endowed with a group structure such that the operations of multiplication and inversion are analytic. Examples of real Lie groups are $GL(n,\mathbb{R})$, $O(n)$, $U(n)$, $\mathbb{R}^n$, $\mathbb{T}^m$. $GL(n,\mathbb{C})$ is a complex Lie group. Every complex Lie group can obviously be regarded also as a real Lie group.

We mention that in the definition of a Lie group the requirement that the underlying manifold be analytic is superfluous, i.e., a Lie group may be defined as a topological group which is just a manifold. This fact is however far from trivial, and it is the content of Hilbert's 5th problem. The first successes on the path to its solution were achieved by J. von Neumann and L. S. Pontryagin in the beginning of the thirties. The final result was obtained by A. Gleason, D. Montgomery, and L. Zippin in 1952.

The study of Lie groups originates in the seventh decade of the last century, when Sophus Lie set out to build the theory that now bears its name. [The theory of Lie groups goes beyond the framework of this book. For a first acquaintance with this rather rich and profound subject we refer the reader to the book of M. M. Postinkov [39].] The general theory of topological groups emerged only later (O. Schreier, 1925). Moments of primordial importance in its subsequent development were the publication of the monographs of L. S. Pontryagin (1938) and A. Weil (1940).

Topological groups form a category in which the morphisms are the continuous homeomorphisms. Accordingly, two topological groups G_1 and G_2 are isomorphic if there exists an algebraic (group) isomorphism $G_1 \to G_2$ which is also a homeomorphism (a *topological isomorphism*). In this case we shall write $G_1 \approx G_2$ (the notation $\approx$ will be also used for isomorphism in other categories).

2°. To construct a general theory of topological groups we must first of all answer the following question : are there, on an arbitrary group, "enough" continuous functions, in the sense that they separate points ? It turns out that the answer is affirmative ; moreover, the separating function ϕ can be choosen to be bounded and *right uniformly continuous*. The latter means that for every $\varepsilon > 0$ there is a neighborhood N of the identity element e such that $|\phi(g_1) - \phi(g_2)| < \varepsilon$ whenever $g_1^{-1}g_2 \in N$ (the notions of left and two-sided uniform continuity are defined in similar manner).

THEOREM (G. Birkhoff, S. Kakutani, 1936). *Let* G *be a topological group. Then for any pair* $g,h \in G$, $g \neq h$, *there is a bounded right uniformly continuous function* ϕ *on* G *such that* $\phi(g) \neq \phi(h)$.

PROOF. We may assume, with no loss of generality, that $h = e$; indeed, if ψ is a function in the required class such that $\psi(g^{-1}h) \neq \psi(e)$, then $\phi(s) = \psi(g^{-1}s)$ belongs to the same class and $\phi(g) \neq \phi(h)$.

Let Δ_0 be a symmetric neighborhood of e such that $g \in \Delta_0$. We define by induction a sequence $\{\Delta_k\}_0^\infty$ of neighborhoods of e with the property that $\Delta_{k+1}^2 \subset \Delta_k$. This is possible since $e^2 = e \in \Delta_k$ and multiplication is continuous. Let $\rho \in (0,1)$ be an arbitrary dyadic rational number, i.e., $\rho = 2^{-\ell_1} + \ldots + 2^{-\ell_n}$, where $\ell_1 < \ldots < \ell_n$ are positive integers. Consider the neighborhood $N_\rho = \Delta_{\ell_1} \ldots \Delta_{\ell_n}$ of e. If $\rho \geqslant 1$ we put $N_\rho = G$. We can now define the sought-for function ϕ by the rule $\phi(s) = \inf\{\rho \mid s \in N_\rho\}$. Obviously, ϕ is bounded, and in fact $0 \leqslant \phi(s) \leqslant 1$. Next, $\phi(e) = 0$, since $e \in N_\rho$ for all $\rho > 0$. Now notice that if $\rho < 1$, then $N_\rho \subset \Delta_1 \ldots \Delta_n \subset$ $\subset \Delta_1 \ldots \Delta_{n-2} \Delta_{n-1}^2 \subset \ldots \subset \Delta_1^2$, whence $N_\rho \subset \Delta_0$. Consequently, $g \in N_\rho$ whenever $\rho > 1$, and so $\phi(g) = 1$. It remains to verify that the function ϕ is right uniformly continuous. To this end we show that $N_\rho \subset N_\sigma$ whenever $\rho < \sigma$. Let $\sigma = 2^{-m_1} + \ldots + 2^{-m_p}$, where $m_1 \leqslant \ell_1$, and if $m_1 = \ell_1$, then $m_2 \leqslant \ell_2$, and so forth (i.e., the collection $(m_1, \ldots, m_p)$ lexicographically precedes $(\ell_1, \ldots, \ell_n)$). Discarding in the products N_ρ and N_σ the initial strings that coincide, we may assume from the very beginning that $m_1 < \ell_1$. Then, as above

$$N_\rho \subset \Delta_{\ell_1} \Delta_{\ell_1+1} \ldots \Delta_{\ell_1+n} \subset \Delta_{\ell_1} \Delta_{\ell_1+1} \ldots \Delta_{\ell_1+n-1}^2 \subset$$

$$\subset \Delta_{\ell_1}^2 \subset \Delta_{\ell_1-1} \subset \Delta_{m_1} \subset N_\sigma .$$

Next we establish the inclusion $N_\rho \Delta_k \subset N_{\rho+2^{-k}}$ for all ρ and all $k \geqslant 3$. In some cases it is obvious : for instance, if $\rho + 2^{-(k-2)} \geqslant 1$, or if $\rho < 1$, $k > \ell_n$ (in the last case $N_\rho \Delta_k = N_{\rho+2^{-k}}$). Suppose $\rho + 2^{-(k-2)} < 1$, $k \leqslant \ell_n$, and let ν be the smallest number for which $k \leqslant \ell_\nu$. Then $\rho < r$, where $r = 2^{-\ell_1} + \ldots + 2^{-\ell_{\nu-1}} + 2^{-(\ell_\nu - 1)}$, and hence $N_\rho \Delta_k \subset N_r \Delta_k = N_{r+2^{-k}}$. Since, on the other hand, $r < \rho + 2^{-(\ell_\nu - 1)} \leqslant \rho + 2^{-(k-1)}$, it follows that $r + 2^{-k} < \rho + 2^{-(k-2)}$, which in turn yields the desired inclusion.

Finally, suppose $g_1^{-1}g_2 \in \Delta_k$. Pick any ρ such that $g_1 \in N_\rho$. Then $g_2 \in N_\rho \Delta_k \subset N_{\rho+2^{-(k-2)}}$. By construction, $\phi(g_2) \leqslant \rho + 2^{-(k-2)}$. Minimizing ρ we get $\phi(g_2) \leqslant \phi(g_1) + 2^{-(k-2)}$. The roles of g_1 and g_2 can be interchanged thanks to the symmetry of the neighborhood Δ_k. We conclude that $|\phi(g_1) - \phi(g_2)| \leqslant 2^{-(k-2)}$ whenever $g_1^{-1}g_2 \in \Delta_k$, i.e., ϕ is right uniformly continuous.

□

En route we established also the following result.

THEOREM. *Every topological space endowed with a structure of topological group is completely regular.*

PROOF. Given any point h and any neighborhood N of h we must produce a "bump function", i.e., a continuous function ψ such that $\psi(h) = 1$ and $\psi(g) = 0$ for all $g \notin N$. Again, with no loss of generality, we may assume that $h = e$. In the preceding proof $\phi(e) = 0$ and $\phi(g) = 1$ for all $g \notin \Delta_0$. Hence, it suffices to take $\Delta_0 \subset N$ and put $\psi = 1 - \phi$.

□

One is naturally led to asking whether every space endowed with a structure of topological group is normal. The negative answer to this question given by A. A. Markov in 1941 relies on certain indirect considerations. We can however give the following explicit example.

Example. Consider the additive group of all real-valued functions on the segment $I = [0,1]$, endowed with the topology of point convergence, or, equivalently, the topological power $\mathbb{R}^I$. *This is a topological group whose underlying space is not normal* (in general, $\mathbb{R}^\Omega$ is not normal whenever Ω is not countable).

3°. On every function ϕ on the group G one can act by *right translation* : $(R(h)\phi)(g) = \phi(gh)$, or by *left translation* : $(L(h)\phi)(g) = \phi(h^{-1}g)$. These are very important and systematically used operations. Obviously, $R(h)$ and $L(h)$ are invertible linear operators in the space $F(G)$ of all functions on G [$F(G)$ can be endowed (if one wishes so) with the topology of pointwise convergence ; it is not a Banach space if $|G|$ is not finite]. Moreover, $R(h_1h_2) = R(h_1)R(h_2)$, $R(e) = E$, and similarly for $L(h)$. A subspace $\Phi \subset F(G)$ is said to be *right (left) invariant* if it is invariant under all operators $R(h)$ (respectively, $L(h)$). A subspace which is both right and left invariant is called *two-sided* or *bi-invariant.* Examples of biinvariant subspaces are the spaces $B(G)$ of bounded functions, $C(G)$ of continuous functions, and $UC_r(G)$ ($UC_\ell(G)$, $UC(G)$) of right (respectively, left, two-sided) uniformly continuous functions. The intersection of any of the indicated spaces with $B(G)$ will be denoted by appending the letter B to the original notation. Given an arbitrary function $\phi \in F(G)$, the linear span of all its right translates $R(h)\phi$ is right invariant ; it is the smallest right invariant subspace containing ϕ. The same holds true for left translates.

Example. Consider the group of all matrices of the form $\begin{pmatrix} \alpha & \beta \\ 0 & 1 \end{pmatrix}$, $\alpha > 0$, $\beta \in \mathbb{R}$, with the usual multiplication and topology. The linear span of the right translates of the function β consists of all functions of the form $a\alpha + b\beta$ with a,b $\mathbb{R}$. The linear span of the left translates of β consists of all functions $c\beta + d$ with $c,d \in \mathbb{R}$.

2. TOPOLOGICAL SEMIGROUPS

A *topological semigroup* is a semigroup endowed with a topology relative to which the multiplication operation is continuous. Examples : any topological group ; any subsemigroup of a topological group with the induced topology, in particular, the additive semigroups $\mathbb{Z}_+ = \{k \mid k \in \mathbb{Z},\ k \geq 0\}$ and $\mathbb{R}_+ = \{t \mid t \in \mathbb{R},\ t \geq 0\}$, the multiplicative semigroup End B, and the multiplicative semigroup of any Banach algebra. We should emphasize that a topological semigroup which algebraically is a group is not necessarily a topological group.

Example. Consider the additive group of all sequences of real numbers $\xi = \{\xi_k\}_1^\infty$, with termwise addition. For arbitrarily given positive numbers α and ε, declare as neighborhood of the point ξ the set of all points η such that $-\alpha k^{-1} < \eta_k - \xi_k < \varepsilon$, $k = 1,2,3,\ldots$. In the topology defined by this family of neighborhoods addition is continuous, whereas inversion is discontinuous. The topology in question is obviously Hausdorff.

THEOREM. *Let* S *be a compact semigroup with identity* e. *The group* S' *of all invertible elements of* S *with the induced topology is a compact group.*

PROOF. In $S \times S$ consider the set Γ of all pairs (s,t) such that $st = e$. It is closed, being the preimage of e under the continuous map $(s,t) \to st$. Hence, Γ is compact. But S' is the image of Γ under the projection $(s,t) \to s$, and since the latter is continuous, S' is compact. It remains to show that *in a compact semigroup which algebraically is a group the inversion operation is continuous.*

Suppose inversion is discontinuous at the point s. Then there exists a neighborhood M of s^{-1} with the property that in any neighborhood N of s one can find a point s_N such that $s_N^{-1} \notin M$. But the complement M^c of M in S is compact, and so the net $\{s_N^{-1}\}$ has a limit point $t \in M^c$. Since $st \neq e$, the continuity of multiplication guarantees the existence of neighborhoods N_0 and M_0 of s and t such that $e \notin N_0M_0$. At the same time, there is a neighborhood $N \subset N_0$ such that $s_N^{-1} \in M_0$, whence $e = s_Ns_N^{-1} \in N_0M_0$: contradiction.

□

To illustrate this theorem, consider the unit disk $\mathbb{D} = \{\lambda \mid \lambda \in \mathbb{C}, |\lambda| \leqslant 1\}$. It is a compact multiplicative semigroup, and its group of invertible elements is $\mathbb{D}' = \mathbb{T}$.

We remark here that every Hausdorff topological space can be endowed with a structure of topological semigroup by choosing an arbitrary point 0 and setting st = 0 for all s and t. In this respect groups differ sharply from general semigroups.

Exercise. *A segment cannot be endowed with a structure of topological group.*

A profound connection exists between compact semigroups and compact groups. It was discovered by A. K. Sushkevich (Suschkewitsch, 1928), and then generalized by D. Rees (1940) and A. H. Clifford (1948) in the algebraic direction, and by K. Numakura (1952) in the topological direction. [The algebraic theory of semigroups is treated in the monograph with the same title by A. Clifford and G. Preston [7] .]

Let us define the main object that will intervene in the ensuing discussion. The *Sushkevich kernel* (*"Kerngruppe"*, or simply *kernel*) of a semigroup is a minimal two-sided ideal. Not every semigroup has a kernel. For example, the additive semigroup $\mathbb{Z}_+ = \{0,1,2,...\}$ does not have a kernel, since its ideals are $\{m \mid m \geq n\}$, $n \in \mathbb{Z}_+$, and none of them is minimal. If a semigroup contains a null element 0, i.e., $s0 = 0s = 0$ for all s, then obviously the set $\{0\}$ is its kernel. A fundamental result of the investigations of the authors mentioned above is the following

KERNEL THEOREM. *Every compact semigroup* S *possesses a Sushkevich kernel* K. *It enjoys the following properties* :

1) K *is closed ;*

2) if $\{J_\alpha\}$ *and* $\{I_\beta\}$ *are the sets of all left and respectively right minimal ideals of* S, *then*

$$\boxed{K = \bigcup_\alpha J_\alpha = \bigcup_\beta I_\beta = \bigcup_{\alpha,\beta} G_{\beta\alpha}}$$

where $G_{\beta\alpha} = I_\beta \cap J_\alpha$ *are compact subgroups of* S *which are mutually topologically isomorphic ;*

3) $G_{\beta\alpha}G_{\delta\gamma} = G_{\beta\gamma}$.

[It is clear that the ideals in the family $\{J_\alpha\}$ are pairwise disjoint, as are those in the family $\{I_\beta\}$. Consequently, *the groups* $G_{\beta\alpha}$ *are also pairwise disjoint.*]

PROOF. Since multiplication by an arbitrarily given element s is continuous and S is compact, the left principal ideal Ss generated by s is closed. Hence, every left ideal contains a closed ideal. Consequently, the set of all minimal left ideals coincides with the set of all closed minimal left ideals, and the latter is not empty by Zorn's Lemma (which applies thanks to the compactness of S). An analogous assertion is valid for right ideals.

Now pick an arbitrary right (left) ideal I (respectively, J). Then $IJ \subset I \cap J$, and so $I \cap J \neq \emptyset$. If I is a two-sided ideal, then $I \cap J$ is a left ideal contained in J. If, in addition, J is a minimal left ideal, then $I \cap J = J$, i.e., $I \supset J$. It follows that the intersection of all two-sided ideals contains J, and so is not empty. This intersection is the smallest two-sided ideal, that is, the Sushkevich kernel K of S.

Next, the set $StS \subset K$ is a two-sided ideal for every $t \in K$, and so $K = StS$. This shows that K is compact, and hence closed.

We already showed that K contains all minimal left ideal. Let J be a minimal left ideal. Then the left ideal Jt is minimal for every $t \in S$, since it is generated by any of its elements. In fact, the elements of Jt have the form st (with $s \in J$), and $S(st) = (Ss)t = Jt$, because $Ss = J$ by the minimality of J. It follows that the union Σ of all minimal left ideals is a right ideal. Also, Σ is trivially a left ideal, and so a two-sided ideal. Therefore, $\Sigma \supset K$, which finally gives $\Sigma = K$. In a similar manner, the union of all minimal right ideals coincides with K.

Let us show that if J (I) is a minimal left (respectively, right) ideal, then $G = I \cap J$ is a compact group. That $G \neq \emptyset$ follows, as we remarked earlier, from the inclusion $G \supset IJ$. Since I and J are closed, and hence compact, G is compact too. By the preceding theorem, it suffices to verify that G is an abstract group. That G is a semigroup is plain : $GG \subset IJ \subset G$.

We show that divisibility at right (for definiteness) holds in G. Let s, t G. Since $Jt \subset J$ and Jt is a left ideal, we have $Jt = J$. Hence, there is an $u \in J$ such that $ut = s$. Let us show that $u \in I$. Suppose $u \notin I$. Then u, as an element of the kernel K ($u \in J \subset K$!), belongs to a minimal right ideal $I_1 \neq I$, and $I_1 \cap I = \emptyset$. But then $s = ut \in I_1$, whence $u \notin I$, which contradicts the fact that from the very beginning $u \in G = I \cap J$.

The identity element of the group G is an idempotent. Therefore, *every compact semigroup contains an idempotent* (this intermediary result is important in its own right).

Finally, let us prove 3). Let $s \in G_{\beta\alpha}$ and $t \in G_{\delta\gamma}$. Then $st \in G_{\beta\gamma}$, since s belongs to the right ideal I_β and t to the left ideal J_γ. Consequently, $G_{\beta\alpha}G_{\delta\gamma} \subset G_{\beta\gamma}$. [Notice that this already means that the partition $K = \cup_{\alpha,\beta} G_{\beta\alpha}$ is stable and that equality 3) holds if one regards it as a multiplication table for a quotient semigroup. A semigroup with such a multiplication table is called a *matrix band*. If it arises, as in the present case, as a quotient semigroup, then the original semigroup is called a *matrix band of classes of a partition*. Thus, *the Sushkevich kernel of any compact semigroup is a matrix band of compact groups*.] Furthermore, the product $\Gamma = G_{\beta\alpha}G_{\delta\gamma}$ is a compact semigroup : its compactness follows from the compactness of the factors $G_{\beta\alpha}$ and $G_{\delta\gamma}$, and we have $\Gamma^2 = G_{\beta\alpha}(G_{\delta\gamma}G_{\beta\alpha})G_{\delta\gamma} \subset G_{\beta\alpha}G_{\delta\alpha}G_{\delta\gamma} \subset G_{\beta\alpha}G_{\delta\gamma} = \Gamma$. Therefore, Γ contains an idempotent, and since $\Gamma \subset G_{\beta\delta}$, the latter must coincide with the identity $e_{\beta\gamma}$ of the group $G_{\beta\gamma}$. By the definition of Γ, we can write $e_{\beta\gamma} = uv$, with $u \in G_{\beta\alpha}$ and $v \in G_{\delta\gamma}$. Now let $w \in G_{\beta\gamma}$. Then $w = e_{\beta\gamma}w = uvw$. But $vw \in G_{\delta\gamma}G_{\beta\gamma} \subset G_{\delta\gamma}$ and $u \in G_{\beta\alpha}$. It follows that $w \in G_{\beta\alpha}G_{\delta\gamma} = \Gamma$. We thus proved the opposite inclusion $G_{\beta\gamma} \subset G_{\beta\alpha}G_{\delta\gamma}$, and hence equality 3).

It remains to show that the groups $G_{\beta\alpha}$ are all isomorphic. Pick one of these groups, say G_{00}. Let e_{00} be the identity element of G_{00}. Since $G_{0\alpha}G_{0\beta} = G_{00}$, there exist $u_\alpha \in G_{0\alpha}$ and $v_\beta \in G_{0\beta}$ such that $e_{00} = u_\alpha v_\beta$. Consider the map $i_{\beta\alpha}: s \to v_\beta s u_\alpha$, which in view of 3) acts from G_{00} to $G_{\beta\alpha}$. It is a homomorphism:

$$i_{\beta\alpha}s \cdot i_{\beta\alpha}t = v_\beta s(u_\alpha v_\beta)tu_\alpha = v_\beta(se_{00})u_\alpha = v_\beta(st)u_\alpha = i_{\beta\alpha}(st).$$

Consequently, $e_{\beta\alpha} = v_\beta e_{00} u_\alpha$ is the identity element of $G_{\beta\alpha}$. The map $j_{\beta\alpha}: G_{\beta\alpha} \to G_{00}$ defined as $j_{\beta\alpha} z = u_\alpha z v_\beta$ is a two-sided inverse of $i_{\beta\alpha}$. In fact,

$$j_{\beta\alpha} i_{\beta\alpha} s = u_\alpha (v_\beta s u_\alpha) v_\beta = e_{00} s e_{00} = s ,$$

and

$$i_{\beta\alpha} j_{\beta\alpha} z = v_\beta (u_\alpha z v_\beta) u_\alpha = v_\beta (e_{00} u_\alpha z v_\beta e_{00}) u_\alpha = e_{\beta\alpha} z e_{\beta\alpha} = z.$$

Thus, $i_{\beta\alpha}$ is an isomorphism. It is obviously continuous, and so is its inverse, i.e., $i_{\beta\alpha}$ is a topological isomorphism. This completes the proof of the theorem.

□

Exercise 1. $I_\beta J_\alpha = G_{\beta\alpha}$ *and* $J_\alpha I_\beta = K$.

Exercise 2. *Let* S *be a finite semigroup with one generator* g. *Then* $S = \{g, \ldots, g^{d-1}, g^d, \ldots, g^{d+p-1}\}$, *where* $g^{d+p} = g^d$. *The numbers* p *and* d *are called the period and respectively the index of* S *(and* $d-1$ *is sometimes called the preperiod of* S*).* $K = \{g^d, \ldots, g^{d+p-1}\}$ *is a subgroup of* S *(it is precisely the Sushkevich kernel of* S*). The unique idempotent of* S *is the identity element* e *of* K. *It is given by* $e = g^k$, *where* $d \leqslant k \leqslant d + p - 1$, $k \equiv 0 \pmod{p}$. *The group* K *is cyclic and* g^{k+1} *is a generator of* K (G. Frobenius, 1895).

Let S be an arbitrary compact semigroup. We let ℓ and r denote the cardinality of the set of all minimal and respectively right ideals of S. We say in this case that S is of *type* $\ell \times r$. If ℓ and r are finite we say that S is of *finite type*. The Kernel Theorem has the following consequences.

COROLLARY 1. *The Sushkevich kernel of a compact semigroup* S *is a group (and hence a compact group) if and only if* S *is of type* 1×1.

In fact, if S is of type 1×1 its kernel is equal to the group G_{00}, whereas if S is of any other type the kernel is not a group, since it contains more than one idempotent. □

COROLLARY 2. *The Sushkevich kernel* K *of a compact Abelian semigroup* S *is a compact Abelian group.*

In fact, any minimal left of right ideal in S is two-sided and hence contains K. Therefore, it is unique and S is of type 1×1. □

The kernel K of an arbitrary compact semigroup S can be represented as $K = StS$ for any $t \in K$. Consequently, *the kernel of any compact connected semigroup is connected.* We see that the Kernel Theorem admits also the following

COROLLARY 3. *Every compact connected semigroup* S *of finite type is necessarily of type* 1×1.

In fact, the kernel of such an S is the union of a finite number of pairwise disjoint compact sets $G_{\beta\alpha}$ and at the same time is connected.

□

Now let S be an arbitrary compact semigroup of type 1×1 and let e denote the identity element of its kernel K. The following lemma will prove useful in what follows.

LEMMA. e *belongs to the center of* S. *The map* $\hat{e}$ *given by* $\hat{e}s = se$ *is a retraction of* S *onto* K *(i.e.,* $\hat{e}S = K$ *and* $\hat{e}|K = \mathrm{id}$) *and a homomorphism.*

PROOF. se and es belong to K for every $s \in S$, and so $se = ese = es$. That $\hat{e}$ is a retraction is obvious. Finally, $(st)e = (se)(te)$ because $ete = te$.

□

We call $\hat{e}$ the *canonical retraction of* S *onto* K.

COROLLARY. *Every closed subsemigroup* S *of a compact group* G *is a subgroup (and thus a compact subgroup).*

In fact, the identity element e of the kernel K of S is an idempotent in G, and so it coincides with the identity element of G. But then $\hat{e} = \mathrm{id}$, i.e., $S = K$. Since a group has only one idempotent, S is of type 1×1, i.e., K is a group, and hence so is S.

□

To conclude this section we mention some differences between the semigroup and group case which arise when function spaces are considered. [Henceforth by semigroup we mean "topological semigroup". Topological semigroups form a category in which the morphisms are the continuous homomorphisms. The notion of *topological isomorphism* of semigroups is readily defined (as in the case of groups).]

Let S be a semigroup. The *left translation* (acting on functions) on S is defined as $(\Lambda(s)\phi)(t) = \phi(st)$. This definition is imposed by the noninvertibility of the elements $s \in S$, and the cost of this modification is that the order of the factors in the formula $L(h_1h_2) = L(h_1)L(h_2)$, valid for groups, changes to $\Lambda(s_1s_2) = \Lambda(s_2)\Lambda(s_1)$. The notion of (say, right) uniform continuity for functions on S must also be modified to account for noninvertibility of elements. Suppose the function ϕ is bounded and continuous. We say that ϕ is *right uniformly continuous* if $\lim_{t \to s} \sup_u |\phi(ut) - \phi(us)| = 0$ for all $s \in S$. If S is a group we recover the previous definition. If S has an identity element, it is not necessary to require continuity beforehand. The Banach space of all bounded right uniformly continuous functions on S will be denoted by $UCB_r(S)$. It is biinvariant. [As a rule, we shall use for function spaces on semigroups the same notations as for the analogous spaces on groups.]

3. INVARIANT MEASURES AND MEANS

1°. We remaind the reader that a map $f : S \to S$ of a space S with a measure $\sigma \equiv ds$ is said to be *measure preserving* if the preimage $f^{-1}M$ is measurable and $\sigma(f^{-1}M) = \sigma(M)$ for every measurable set M S. Let f act on functions on S by the rule $(f*\phi)(s) = \phi(fs)$. If f is measure preserving, then the integral $\int \phi ds$ over S is invariant under f for all functions $\phi \in L_1(S, ds)$: $\int (f*\phi) ds = \int \phi ds$.

Let S be a semigroup. A measure $\sigma \neq 0$ on S is said to be *right invariant* if it is invariant under all right translations $s \to st$ with $t \in S$. *Left invariance* and *two-sided invariance, or biinvariance* are defined in the same manner (the term "biinvariant" will be usually omitted, and sometimes we will do the same with "one-sided invariance", if there is no danger of confusion).

Example. Let G be a group. We define a measure dg on all subsets of G by the rule $M \to |M|$ (= the number of elements in M is M is finite, and ∞ otherwise). The measure dg is obviously invariant. The space of dg-integrable functions on G, denoted $\ell_1(G)$, consists of all functions ϕ such that $||\phi|| = \sum_g |\phi(g)| < \infty$. Clearly, $\int \phi \, dg = \sum_g \phi(g)$. If G is finite, it is convenient to normalize dg, setting $\int_M dg = |M|/|G|$, so that $\int_G dg = 1$. In this case $\int_G \phi \, dg = |G|^{-1} \sum_g \phi(g)$ is the *mean (average) of* ϕ *over* G.

The general problem of the existence of an invariant measure on locally compact groups was solved by A. Haar in 1933.

HAAR'S THEOREM. *On every locally compact group there exists a right invariant (left invariant) regular Borel measure.*

□

The proof of this fundamental result can be found, for example, in P. Halmos's book [17]. Haar's original proof required a supplementary condition (that the group be second countable) which was later removed by A. Weil. The "left" version of Haar's Theorem follows from the "right" version since the measure σ^*, $\sigma^*(M) = \sigma(M^{-1})$, is left invariant whenever σ is right invariant. A biinvariant measure does not necessarily exist. A measure on a locally compact group with the properties indicated in Haar's Theorem is called a *righ (left) Haar measure*. A measure that is both a right and left Haar measure is referred to as a (biinvariant) *Haar measure.*

Example. *The Lebesgue measure on* $\mathbb{R}^n$ *is a Haar measure.*

The local compactness constraint on the group cannot be dropped if the measure is to enjoy nice topological properties.

Example. *On an infinite dimensional Banach space* B, *regarded as an additive group, there is no invariant measure taking finite positive values on every ball.* In fact, there exists a sequence $\{e_k\} \subset B$, $\|e_k\| = 1$, such that $\|e_k - e_j\| \geqslant 1/2$ for all $j \neq k$. The balls $\|x - e_k\| < 1/4$, $k = 1,2,3,\ldots$ are pairwise disjoint, have all the same measure, and at the same time are all"packed" in the ball $\|x\| < 5/4$.

We return to the case of locally compact groups. It is clear that if dg is a right (left) Haar measure, then such is every proportional measure cdg, with c = const. Do we obtain in this way all the right (respectively, left) Haar measures ? This question was answered affirmatively by J. von Neumann in 1936.

VON NEUMANN'S THEOREM. *All right (left) Haar measures on a locally compact group are proportional.*

□

In his work von Neumann assumed, as Haar did, that the group is second countable, and again this requirement was later removed by A. Weil. The proof of von Neumann's Theorem can be found in Halmos's book [17] .

We next give an example of a group on which there is no biinvariant Haar measure. To this end it suffices to exhibit a locally compact group on which a right Haar measure is not proportional to a left one.

Example. Consider the group (already encountered in Sec.1) of all matrices of the form $\begin{pmatrix} \alpha & \beta \\ 0 & 1 \end{pmatrix}$ with $\alpha > 0$ and $\beta \in \mathbb{R}$.

Set $\tau(M) = \iint_M \alpha^{-2}\, d\alpha d\beta$ for every Borel set M. It is readily verified that this defines a left Haar measure τ. At the same time, $\sigma(M) = \iint_M \alpha^{-1}\, d\alpha d\beta$ is a right Haar measure which is not proportional to $\tau(M)$.

In the next exercises σ is a right (or left) Haar measure on a locally compact group G.

Exercise 1. $\sigma(D) > 0$ *for every open set* D.

Exercise 2. G *is discrete if and only if* $\sigma(\{e\}) > 0$.

Exercise 3. G *is compact if and only if* $\sigma(G) < \infty$.

Let $\sigma = dh$ be a right Haar measure on a locally compact group G. For an arbitrary element $g \in G$ we put $\sigma_g(M) = \sigma(gM)$ for every Borel set M. Since σ_g is again a right Haar measure, it is proportional to σ : $\sigma_g = \Delta(g)\sigma$.

THEOREM. *The map* $\Delta : G \to \mathbb{R}_+$ *is a continuous homomorphism of* G *into the multiplicative group of positive numbers.*

PROOF. It follows from the identity $\sigma(gM) = \Delta(g)\sigma(M)$ and the associativity rule that $\Delta(g_1g_2) = \Delta(g_1)\Delta(g_2)$, i.e., Δ is a homomorphism. To establish the continuity of Δ it suffices to pick an arbitrary continuous function $\phi \geq 0$, $\phi \neq 0$, with compact support and remark that $\Delta(g) = \int \phi(g^{-1}h)\,dh/(\int \phi\, dh)$.

□

Consider next the measure defined by the formula $\tau(M) = \int_M \Delta(h^{-1})\,dh$. We have :

$$\tau(gM) = \int_{gM} \Delta(h^{-1})\,dh = \int_M \Delta(h^{-1}g^{-1})\,d_h(gh) =$$

$$= \Delta(g^{-1})\Delta(g)\int_M \Delta(h^{-1})\,dh = \tau(M),$$

i.e., τ is a left Haar measure. We thus obtain the following

THEOREM. *In order for a given right Haar measure on a locally compact group* G *to be biinvariant (or, equivalently, in order for a biinvariant Haar measure to exist on* G*) it is necessary and sufficient that* $\Delta(g) = 1$ *for all* $g \in G$.

□

Groups with the last property are called *unimodular*. Obviously, every Abelian locally compact group is unimodular. However, there exist also non-Abelian unimodular groups.

Example 1. *Every compact group is unimodular.* In fact, its image in $\mathbb{R}_+$ under the homomorphism Δ is a compact subgroup, and so $\text{Im}\,\Delta = \{1\}$.

Example 2. *Every discrete group is unimodular.* In fact, $\Delta(g) = \sigma(g\{e\})/\sigma(\{e\}) = \sigma(\{g\})/\sigma(\{e\}) = 1$, since $\{g\} = \{e\}g$ and σ is a right Haar measure.

We remark that every Haar measure on a discrete group is proportional to $\sigma_0(M) = |M|$.

Exercise. *Let* $\sigma \equiv dg$ *be a right Haar measure and let* $\sigma^*(M) = \sigma(M^{-1})$ *be the corresponding left Haar measure. Then* $\sigma^*(M) = \int_M \Delta(g^{-1})dg$. *It follows from this formula that* $\sigma(M^{-1}) > 0$ *if and only if* $\sigma(M) > 0$. *In a unimodular group* $\sigma(M^{-1}) = \sigma(M)$ *for all* M.

2°. The existence of a right (for definiteness) Haar measure dg permits us to consider, on any locally compact group, the Banach spaces of functions $L_p(G)$, $1 \leq p \leq \infty$. *They are all biinvariant.* A special role in this family is played by $L_1(G)$. The reason is that $L_1(G)$ possesses a natural structure of Banach algebra (generally speaking, without unit ; in this case the unit is adjoined formally). The corresponding multiplication operation is called *convolution*. By definition, the *convolution of the functions* $\phi, \psi \in L_1(G)$ is the function

$$(\phi * \psi)(h) = \int \phi(g)\psi(hg^{-1})dg \quad .$$

That this integral exists almost everywhere (a.e.) and belongs to $L_1(G)$ follows from Fubini's Theorem :

$$\int dh \int |\phi(g)||\psi(hg^{-1})|dg = \int|\phi(g)|dg \int |\psi(hg^{-1})|dh =$$

$$= \int|\phi(g)|dg \int|\psi(h)|dh < \infty.$$

This shows also that $||\phi * \psi|| \leq ||\phi|| \, ||\psi||$. It is easily seen that the convolution enjoys all the properties of a multiplication, i.e., it is bilinear (which is obvious) and associative. Generally speaking, however, it is not commutative.

It is readily verified that *the center of the algebra* $L_1(G)$ *consists of the functions* ϕ *that satisfy the equation* $\phi(h^{-1}gh) = \phi(g)$ *(all* $h,g \in G$*)*. Such functions are termed *central*. *The convolution operation in* $L_1(G)$ *is commutative if and only if* G *is Abelian.* The Banach algebra $L_1(G)$ is called the *group algebra* of the locally compact group G.

Exercise. *The group algebra* $L_1(G)$ *possesses a unit if and only if* G *is discrete.*

A more general construction of a group algebra presumes that on the locally compact group G there is given a locally bounded measurable function $\alpha(g) > 0$ which satisfies the *ring condition* : $\alpha(g_1g_2) \leq \alpha(g_1)\alpha(g_2)$. Such a function is called a *weight* and defines the Banach algebra $L_1(G;\alpha)$ consisting of all functions ϕ on G for which the norm $\int|\phi(g)|\alpha(g)dg$ is finite ; the multiplication in $L_1(G;\alpha)$ is, as above, the convolution. The inequality $||\phi * \psi|| \leq ||\phi|| \, ||\psi||$ is guaranteed by the ring property of the weight.

The convolution leads, in particular, to an important construction known as a sliding mean. To describe it we remark that the convolution $\phi * \psi$ is well defined (in the usual manner) even if the function ψ is merely locally summable (i.e., summable over every compact set), provided that $\phi \in L_1(G)$ has compact support. Now consider the directed set of all precompact neighborhoods of the identity element $e \in G$. To each such neighborhood N we correspond a function $\mu_N \in L_1(G)$ with the following properties : $\mu_N \geq 0$, $\mu_N(g) = 0$ if $g \notin N$, and $\int \mu_N dg = 1$ (and also μ_N continuous, if one wishes so). The convolution $\psi_N = \mu_N * \psi$ is called the *sliding* (or *moving*) *mean with kernel* μ_N for the locally summable function ψ. It is natural to expect that for "small" neighborhoods N the sliding mean ψ_N is close to ψ. [The net $\{\mu_N\}$ is referred to as an *approximate unit* for $L_1(G)$.]

THEOREM. *The net* $\{\psi_N\}$ *converges to* ψ *in the* L_1*-norm on every compact subset of* G. *If* $\psi \in L_1(G)$, *then* $\psi_N \to \psi$ *in* $L_1(G)$. *If* ψ *is continuous, then* ψ_N *is continuous for every* N, *and* $\psi_N \to \psi$ *uniformly on any compact subset of* G.

PROOF. Clearly,

$$(\psi_N - \psi)(h) = \int_N \mu_N(g)[\psi(hg^{-1}) - \psi(h)]dg .$$

Therefore,

$$\|\psi_N - \psi\|_{L_1(Q)} \leq \int_N \mu_N(g)dg \int_Q |\psi(hg^{-1}) - \psi(h)|dh$$

for every compact set $Q \subset G$ (and if $\psi \in L_1(G)$, then for $Q = G$, too). Since ψ is locally (and, for $Q = G$, globally) summable, for every $\varepsilon > 0$ there is a neighborhood N_ε of e such that $\int_Q |\psi(hg^{-1}) - \psi(h)|dh < \varepsilon$ for all $g \in N_\varepsilon$. Consequently, $\|\psi_N - \psi\|_{L_1(Q)} \leq \varepsilon\int_N \mu_N dg = \varepsilon$ whenever $N \subset N_\varepsilon$.

Now suppose ψ is continuous. Then it is uniformly continuous on every compact set. Hence, for every $\varepsilon > 0$ and every compact set Q there is a (precompact) neighborhood N of e such that $|\psi(hg^{-1}) - \psi(h)| < \varepsilon$ for all $h \in Q$ and all $g \in N$. Consequently,

$\sup_Q |\psi_N - \psi| \leqslant \varepsilon$. The continuity of each of the functions ψ_N follows in analogous manner from the formula

$$\psi_N(h_1) - \psi_N(h_2) = \int_N \mu_N(g)[\psi(h_1 g^{-1}) - \psi(h_2 g^{-1})]dg.$$

□

3°. We next consider invariant measures on semigroups. As it turns out, such a measure can be "meager" in the sense that there may exists open sets of null measure. In point of fact, the support of such a measure may even reduce to one point.

Example. Let S be a multiplicative Abelian semigroup with a null element 0. The preimage of any set $M \subset S$ under translation by 0 is equal to S if $0 \in M$ and is empty otherwise. Since there necessarily exist measurable sets containing 0 (the complement of any measurable set is measurable !), every invariant measure $\sigma(M)$ on S is equal to $\sigma(S)\delta_M(0)$, where δ_M denotes the indicator function of M.

In contradistinction to the case of groups, by far not all locally compact (and even not all discrete) semigroups admit an invariant measure.

Example. *On the additive semigroup* $\mathbb{Z}_+$ *there are no invariant measures.* In fact, let σ be such a measure, and let M be a set of positive measure. Denote $M_n = M \cap [0,n]$. Then $\sigma(M_n) > 0$ for sufficiently large n. At the same time, the preimage of M_n under translation by $n+1$ is empty.

Examining this example one can guess that the existence of invariant measures is connected with that of minimal ideals.

THEOREM. *Suppose that on a semigroup there is given a right invariant measure* μ *defined at least on the* σ*-algebra generated by the open sets. Then its support* $\operatorname{supp}\mu$ *is a closed right ideal.*

PROOF. The support of a measure is closed by definition.

Let $s \in \operatorname{supp} \mu$, but $st \in \operatorname{supp} \mu$ for some t. Then there exists a neighborhood N of st such that $\mu(N) = 0$. The preimage M of N under right translation by t is a neighborhood of s (by the continuity of right translation) and $\mu(M) = 0$ (by the right invariance of μ). Hence, $s \notin \operatorname{supp} \mu$: contradiction.

□

For compact semigroups a criterion for the existence of an invariant measure was found by W. G. Rosen (1956) in terms of the Sushkevich kernel.

THEOREM. *Let* S *be a compact semigroup. Then on* S *there exists a right invariant Borel measure* μ *if and only if* S *is of type* $1 \times r$.

PROOF. NECESSITY. Let J be a minimal left ideal in S. Then J is manifestly measurable (being closed, and hence compact). If $t \in J$, then $St \subset J$ and St is a left ideal. Therefore, $St = J$, i.e., the preimage of J under right translation by t is the full semigroup S. By the right invariance of μ, $\mu(S) = \mu(J)$, whence $\mu(S \setminus J) = 0$. Consequently, $\operatorname{supp} \mu \subset J$, and since J is arbitrary, it follows that the intersection of all minimal left ideals is not empty. But then there is a unique minimal left ideal, i.e., S is of type $1 \times r$.

SUFFICIENCY. Suppose S is of type $1 \times r$, i.e., it has a unique minimal left ideal J_0. Consider the family $\{I_\beta\}$ of all minimal right ideals of S. By the Kernel Theorem, the Sushkevich kernel K of S can be written as $K = \cup_\beta G_{\beta 0}$, where $G_{\beta 0} = I_\beta \cap J_0$ are compact groups. Pick one of these groups, say G_{00}, and let μ_0 be a Haar measure on G_{00}. Now for each Borel set $M \subset S$ put $\mu(M) = \mu_0(M \cap G_{00})$. This defines a Borel measure μ on S. We show that μ is right invariant. Let M_t denote the preimage of M under right translation by t ; M_t is a Borel set by the continuity of translation. We claim that $\mu(M_t) = \mu(M)$. In fact, $\mu(M_t) = \mu_0(M_t \cap G_{00})$, by definition. But $M_t \cap G_{00} = \{s \mid s \in G_{00},\ st \in M\}$. Since $st = se_{00}t$ for all $s \in G_{00}$ (where e_{00} denotes the identity element of G_{00}), and since

$e_{00}t \in K$ (as K is a two-sided ideal), we may assume from the beginning that $t \in K$. Furthermore, $e_{00}t \in G_{00}$ (by the multiplication table $G_{00}G_{\beta 0} = G_{00}$, which gives $G_{00}K = G_{00}$), and so we may assume also that $t \in G_{00}$. Then $st \in G_{00}$ for all $s \in G_{00}$. Consequently, $M_t \cap G_{00} = \{s \mid s \in G_{00}, st \in M \cap G_{00}\}$, which means that $M_t \cap G_{00}$ is the preimage of $M \cap G_{00}$ under translation by $t \in G_{00}$. By the invariance of the Haar measure μ_0 on G_{00}, $\mu_0(M_t \cap G_{00}) = \mu_0(M \cap G_{00})$, i.e., $\mu(M_t) = \mu(M)$, as claimed.

□

In the proof of the necessity part of the theorem we showed that $\operatorname{supp} \mu \subset K$ for every right invariant Borel measure μ on the compact semigroup S, i.e., *all such measures are supported on the Sushkevich kernel of* S. In the proof of the sufficiency part it turned out that the right invariant measure μ is not only Borelian, but also regular, thanks to the regularity of the Haar measure μ_0. The same proof makes it evident that *for the uniqueness (to within a constant factor) of the right invariant measure* μ *on the compact semigroup* S *it is necessary that* S *be of type* 1×1, *i.e., that the Sushkevick kernel* K *of* S *be a group.* This condition is also sufficient : if it is satisfied, then μ is necessarily a Haar measure on K, since $\operatorname{supp} \mu \subset K$. Moreover, μ turns out to be biinvariant. We thus have the following result.

THEOREM. *Let* S *be a compact semigroup. Then the following are equivalent statements* :

1) S *is of type* 1×1 ;

2) *there exists a biinvariant regular Borel measure on* S;

3) *the right invariant measure on* S *is unique up to a constant factor.*

□

COROLLARY. *On every compact Abelian semigroup there exists a unique (up to a constant factor) invariant regular Borel measure.*

□

4°. Every invariant measure defines an invariant (under the same type of translations) integral. If the measure is normalized then the integral of the constant function $\mathbb{1}$ is equal to 1. On these grounds the corresponding invariant integral may be referred to as an invariant mean. A wider, axiomatic definition of the notion of invariant mean does not require the existence of an invariant measure. Specifically, a *right invariant* (for brevity, *invariant*) *mean* on the semigroup S is an additive homogeneous functional $\langle\cdot\rangle$ on the space $CB(S)$ which is invariant under right translations : $\langle R(s)\phi\rangle = \langle\phi\rangle$ for all $\phi \in CB(S)$ and all $s \in S$, nonnegative : $\langle\phi\rangle \geqslant 0$ whenever $\phi \geqslant 0$, and normalized : $\langle\mathbb{1}\rangle = 1$. Thus, an invariant mean is automatically a (bounded) linear functional on $CB(S)$.

The theory of invariant measures goes back to von Neumann's work (1929).

Exercise 1. *The norm of any invariant mean is equal to one.*

Exercise 2. *Let* $\phi \in CB(S)$ *be real-valued. Then* $\langle\phi\rangle$ *is a real number. This implies that* $\langle\bar{\phi}\rangle = \overline{\langle\phi\rangle}$ *for every* $\phi \in CB(S)$.

Exercise 3. *Let* $\langle\phi\rangle$ *be a right (as we shall always assume) invariant mean on a group. Then* $\langle\phi^*\rangle$*, where* $\phi^*(s) = \phi(s^{-1})$, *is a left invariant mean.*

There are semigroups (and even groups) that admit no invariant means. Even if such a mean exists, it is not necessarily unique.

Exercise. *On a compact group the invariant mean is unique.*

A semigroup on which there exists at least one (right) invariant mean is said to be *(right) amenable*. For groups the prefix "right" or "left" is not necessary, and we simply say "amenable". *Every compact semigroup of type* $1 \times r$ *is right amenable, and every compact group is amenable*, both thanks to the existence of a right invariant measure. For a discrete semigroup S the space $CB(S)$ coincides with the space $B(S)$ of bounded functions. If the topological semigroup S is amenable when endowed with the discrete topology, then it is amenable. In fact, an invariant mean on $B(S)$ retains its properties upon restricting it to $CB(S)$. The following criterion of J. Diximier (1950) proves useful in studying the amenability of semigroups.

THEOREM. *The semigroup* S *is amenable if and only is for every collection of real-valued functions* $\phi_1,\dots,\phi_n \in CB(S)$ *and every collection of elements* $s_1,\dots,s_n \in S$

$$\inf_s \left(\sum_{k=1}^{n} \{ \phi_k(s) - \phi_k(ss_k)\}\right) \leqslant 0 . \tag{1}$$

PROOF. NECESSITY. An invariant mean vanishes on any function of the form $\sum_k \{\phi_k(s) - \phi_k(ss_k)\}$. Consequently, the infimum of such a function is less than or equal to zero.

SUFFICIENCY. In $CB_{\mathbb{R}}(S)$ consider the closed real linear span of the set of all functions of the form $\phi(s) - \phi(st)$. Also, let K denote the set of all functions $\psi \in CB_{\mathbb{R}}(S)$ such that $\inf \psi > 0$. Then K is an open convex cone and, by condition (1), $K \cap L = \emptyset$. By the separating hyperplane theorem, there exists a linear functional $\langle\cdot\rangle$ on $CB_{\mathbb{R}}(S)$ which vanishes identically on the subspace L and is positive on the cone K. Normalizing $\langle\cdot\rangle$ by the condition $\langle \mathbb{1}\rangle = 1$, we obtain an invariant mean on $CB_{\mathbb{R}}(S)$. It remains to extend $\langle\cdot\rangle$ to the complex space $CB(S)$ in the natural manner : $\langle \phi_1 + i\phi_2\rangle = \langle\phi_1\rangle + i\langle\phi_2\rangle$.

□

COROLLARY. *For the amenability of the discrete semigroup (group)* S *it suffices that every finitely-generated subsemigroup (respectively, subgroup) of* S *be amenable.*

PROOF. Let $s_1,\dots,s_n \in S$ and $\phi_1,\dots,\phi_n \in B(S)$. Let Γ denote the subsemigroup (subgroup, if S is a group) of S generated by $s_1,\dots,s_n$. Since Γ is amenable, inequality (1) holds when the infimum is taken over $s \in \Gamma$, and a fortiori when it is taken over $s \in S$. Therefore, S is amenable.

□

On the other hand, we have the following

THEOREM. *Every subgroup* Γ *of a discrete amenable group* G *is amenable.*

PROOF. Consider the left coset space G/Γ and (using the Axiom of Choice) pick a function Θ which assigns to each coset a representative and satisfies $\Theta\Gamma = e$. This permits us to extend every function $\phi \in B(\Gamma)$ to a function $\tilde{\phi} \in B(G)$ by the rule $\tilde{\phi}(g) = \phi(\Theta[g^{-1}]g))$, (where $[h]$ designates the coset of $h \in G$). If now $<\cdot>$ is an invariant mean on G, then $\phi \to <\tilde{\phi}>$ is an invariant mean on Γ (invariance follows from the relation $(R(h)\phi)^{\sim} = R(h)\tilde{\phi}$, $h \in G$).

□

Thus, *the discrete group* G *is amenable if and only if all its finitely-generated subgroups are amenable.* We shall use this criterion to prove von Neumann's theorem (1929) on the amenability of Abelian groups. To this end we need also two preliminary lemmas.

LEMMA 1. *The direct product of two discrete amenable semigroups is amenable.*

PROOF. Let $S = S_1 \times S_2$, and let $<\cdot>_1$ and $<\cdot>_2$ be invariant means on S_1 and S_2, respectively. Then upon averaging every function $\phi(s_1, s_2)$, $\phi \in B(S)$, over $s_2 \in S_2$ we obtain a function $\phi_1(s_1) = <\phi(s_1,\cdot)>_2$, $\phi_1 \in B(S_1)$, which in turn can be averaged over $s_1 \in S_1$. Putting $<\phi> = <\phi_1>_1$ we obtain the sought-for invariant mean on S.

□

LEMMA 2. *Every homomorphic image of a discrete amenable semigroup* Σ *is amenable.*

PROOF. Every homomorphism $h : \Sigma \to S$ induces a Banach space morphism $h^* : B(S) \to B(\Sigma)$. Pick an invariant mean $<\cdot>$ on $B(\Sigma)$ and carry it over to $B(S)$ by the rule : $<\phi>^* = <h^*\phi>$. If h is surjective, i.e., an epimorphism, $<\cdot>^*$ is an invariant mean on S, as it is readily verified.

□

We are now ready to prove the following desirable result (and even in the more general case of semigroups).

THEOREM. *Every Abelian semigroup* S *is amenable.*

PROOF. We may assume that S is discrete, finitely-generated, and possesses an identity element. Let $s_1,\dots,s_k$ be a system of generators of S, i.e., in the additive notation, S is equal to the set of all combinations $\sum_{1\leq i\leq k} n_i s_i$ with $n_1,\dots,n_k \in \mathbb{Z}_+$. We replace $s_1,\dots,s_k$ by formal symbols $\sigma_1,\dots,\sigma_k$ and consider the set Σ of all formal combinations $\sum_{1\leq i\leq k} n_i\sigma_i$. It is a free semigroup with generators $\sigma_1,\dots,\sigma_k$. The mapping

$$\sum_{1\leq i\leq k} n_i\sigma_i \to \sum_{1\leq i\leq k} n_i s_i$$

is clearly an epimorphism $\Sigma \to S$. Hence, by Lemma 2, it suffices to show that Σ is amenable. This is the content of a theorem of Banach, and the corresponding invariant mean is called the *Banach limit*.

THEOREM. *The semigroup* $\mathbb{Z}_+$ *is amenable.*

PROOF. The functions in $B(\mathbb{Z}_+)$ are simply bounded sequences of real numbers $\xi = \{\xi_k\}_0^\infty$. On this Banach space consider the functional

$$p(\xi) = \inf_{m;k_1,\dots,k_m > 0} \{ \overline{\lim_{n\to\infty}} \frac{1}{m} \sum_{i=1}^m \xi_{n+k_i} \} .$$

It is subadditive $(p(\xi + \eta) \leq p(\xi) + p(\eta))$ and positive-homogeneous $(p(\lambda\xi) = \lambda p(\xi) \ \ \forall \lambda > 0)$. The last property is obvious. To prove subadditivity, take $\varepsilon > 0$ and find $k_1,\dots,k_m$ and $\ell_1,\dots,\ell_r$ such that

$$\overline{\lim_{n\to\infty}} \frac{1}{m} \sum_{i=1}^m \xi_{n+k_i} < p(\xi) + \varepsilon$$

and

$$\overline{\lim_{n\to\infty}} \frac{1}{r} \sum_{j=1}^{r} \eta_{n+\ell_j} < p(\eta) + \varepsilon .$$

Then

$$\overline{\lim_{n\to\infty}} \frac{1}{mr} \sum_{i=1}^{m} \sum_{j=1}^{r} (\xi + \eta)_{n+k_i+\ell_j} < p(\xi) + p(\eta) + 2\varepsilon .$$

Consequently, $p(\xi + \eta) < p(\xi) + p(\eta) + 2\varepsilon$, and since ε is arbitrary, $p(\xi + \eta) \leqslant p(\xi) + p(\eta)$.

Now define an additive homogeneous functional $\operatorname{Lim} \xi$ on $B(\mathbb{Z}_+)$ putting first $\operatorname{Lim} \xi = c$ if the sequence ξ is stationary and all $\xi_k = c$, and then extending it to the whole $B(\mathbb{Z}_+)$ by the Hahn-Banach Theorem, preserving the inequality $\operatorname{Lim} \xi \leqslant p(\xi)$ (obviously, for a stationary sequence $p(\xi) = c$). Let us show that $\operatorname{Lim} \xi$ is an invariant mean on $\mathbb{Z}_+$.

Since $p(\xi) \leqslant \sup \xi$, we also have $\operatorname{Lim} \xi \leqslant \sup \xi$. Consequently, $\operatorname{Lim}(-\xi) \leqslant -\inf \xi$, whence $\operatorname{Lim} \xi \geqslant \inf \xi$, and if $\xi \geqslant 0$, then $\operatorname{Lim} \xi \geqslant 0$. It remains to show that $\operatorname{Lim} \xi$ is invariant. Set $\xi'_k = \xi_{k+1}$, $k \in \mathbb{Z}_+$. It suffices to check that $\operatorname{Lim} \xi' = \operatorname{Lim} \xi$. Let $\rho = \xi' - \xi$. Then

$$p(\rho) \leqslant \overline{\lim_{n\to\infty}} \frac{1}{m} \sum_{i=1}^{m} \rho_{n+i-1} = \frac{1}{m} \overline{\lim_{n\to\infty}} (\xi_{n+m} - \xi_n) \leqslant \frac{2}{m} \sup \xi.$$

Letting here $m \to \infty$, we get $p(\rho) \leqslant 0$, whence $\operatorname{Lim} \rho \leqslant 0$. Replacing ξ by $-\xi$, and accordingly ρ by $-\rho$, we obtain $\operatorname{Lim} \rho \geqslant 0$. Therefore, $\operatorname{Lim} \rho = 0$, as needed.

□

Exercise 1. *Let* $\xi, \eta \in B(\mathbb{Z}_+)$ *and suppose that there is an index* ν *such that* $\xi_n = \eta_n$ *for all* $n > \nu$. *Then* $\operatorname{Lim} \xi = \operatorname{Lim} \eta$.

Exercise 2. $\underline{\lim}_{n\to\infty} \xi_n \leqslant \operatorname{Lim} \xi \leqslant \overline{\lim}_{n\to\infty} \xi_n$. *Thus, if the sequence* ξ_n *converges as* $n \to \infty$, *then* $\operatorname{Lim} \xi = \lim_{n\to\infty} \xi_n$. *One can use this equality to define* $\operatorname{Lim} \xi$ *immediately on the space* c *of convergent sequences and then extend it to the whole space* $B(\mathbb{Z}_+)$ *with preservation of the inequality* $\operatorname{Lim} \xi \leqslant p(\xi)$.

Exercise 3. *There exist infinitely many invariant means on*

$\mathbb{Z}_+$ *which coincide with* $\lim_{n\to\infty} \xi_n$ *on the space* c .

Suppose that on the discrete semigroup S there is given a right invariant mean $<\cdot>$. Then the formula $\mu(M) = <\delta_M>$, where δ_M denotes the indicator function of the set M, defines a finitely-additive normalized right invariant measure on S. We use this remark to construct the following example of a nonamenable group.

<u>Example</u> (J. von Neumann, 1929). *The discrete free group* $F(a,b)$ *with two generators* a,b *is not amenable.* To show this we write the elements of $F(a,b)$ (words) in the standard form $W = \ldots b^j a^k$, and we call k the order of the word W. Let W_k denote the set of all words of order k, $k \in \mathbb{Z}$. Then obviously $W_k \cap W_\ell = \emptyset$, $W_k a = W_{k+1}$, and $W_k b \subset W_0$. Consequently, if μ is the measure defined above, then $\mu(W_k)$ does not depend on k, and so $\mu(W_k) = 0$ because μ is finitely-additive and normalized. In particular, $\mu(W_0) = 0$. On the other hand,

$$\mu(W_0) \geqslant \mu(\cup_{k\neq 0} W_k b) \geqslant \mu((F(a,b) \setminus W_0)b) = \mu(F(a,b) \setminus W_0) = 1 ;$$

contradiction.

Thus, *every discrete group which contains a free subgroup with two generators is nonamenable.* It has been a long-standing conjecture that this property is also necessary for the nonamenability of a discrete group. It was disproved by A. Yu. Ol'shanskii (1980), who constructed an example of nonamenable group with no free subgroups.

Further information on invariant means can be found in the book of Greenleaf [16], which discusses the development of the apparatus of invariant means itself, as well as its applications, in particular, to representation theory.

CHAPTER 3

ELEMENTS OF GENERAL REPRESENTATION THEORY

1. ACTIONS AND REPRESENTATIONS

1°. Let G be a topological group and let X be a Hausdorff topological space. We let Homeo X denote the group of all homeomorphisms $X \to X$. An *action of* G *on* X is, by definition, a homomorphism $A : G \to \text{Homeo}\, X$ which is *strongly continuous*, meaning that the map $g \to A(g)x$, $G \to X$, is continuous for every fixed $x \in X$. The range of this map is called the *orbit* of the point x (under the given action A) and is denoted by $O(x)$. The set of all orbits, endowed with the quotient topology, is called the *orbit space* or the *quotient space* of the action A and is denoted by X/A. If it reduces to a point, A is said to be *transitive* and X is called a *homogeneous space* of the group G. We say that two points $x_1, x_2 \in X$ are *equivalent* (under A) if $O(x_1) = O(x_2)$, in which case we write, as is customary, $x_1 \sim x_2$.

Exercise. $x_1 \sim x_2 \Leftrightarrow x_2 \in O(x_1)$, *i.e., the equivalence classes are precisely the orbits.*

Example. Take $X = G$ and set $R(g)h = hg^{-1}$, $L(g)h = gh$. These are the *right* and *left regular actions of the group* G *on itself*. They are obviously transitive. More generally, one can take the space $X = G/\Gamma$ of right (for definiteness) cosets of a closed subgroup Γ in the group G. The rule $L(g)(h\Gamma) = (gh)\Gamma$ defines a transitive action of G on G/Γ.

Let again A be an arbitrary action. Pick an arbitrary point $x \in X$ and set $S_x = \{g \mid A(g)x = x\}$. Obviously, S_x is a closed subgroup of G. It is called the *isotropy* (*stationary*, or *stability*) *subgroup of the point* x or *of the action* A *at* x. The intersection $\bigcap_{x \in X} S_x$ is the *kernel* Ker A of A. The action A is injective or, in the customary terminology, *exact*, if and only if $\operatorname{Ker} A = \{e\}$.

LEMMA. *If* $x_1 \sim x_2$, *then* S_{x_1} *is isomorphic to* S_{x_2}.

PROOF. Let $x_2 = A(h)x_1$. If $A(g)x_1 = x_1$, then $A(hgh^{-1})x_2 = x_2$, i.e., $g \in S_{x_1}$ implies $hgh^{-1} \in S_{x_2}$. In exactly the same manner, if $hgh^{-1} \in S_{x_2}$, then $g \in S_{x_1}$. Therefore, $S_{x_2} = hS_{x_1}h^{-1}$.

□

COROLLARY. *If* X *is a homogeneous space, then all the isotropy subgroups are isomorphic.*

□

Thus, for a transitive action the isotropy subgroup is unique as an abstract group. A transitive action is said to be *effective* if its isotropy subgroup reduces to the identity.

Suppose the group G is compact. Then each orbit $O(x)$ is compact, as the image of G under the continuous map $g \to A(g)x$ into the Hausdorff space X. If A is transitive, then X is necessarily compact.

THEOREM. *Every homogeneous space* X *of a compact group* G *is homeomorphic to the right coset space* X/S_0, *where* S_0 *is the isotropy subgroup. The homeomorphism can be chosen equivariant, meaning that it transforms the action of* G *on* X *into the left action of* G *on* G/S_0.

PROOF. Pick an arbitrary point $x \in X$ and identify S_x and S_0. Denoting the right coset of $g \in G$ by $[g]$, define a map $f : G/S_0 \to X$ by the rule $f[g] = A(g)x$. This definition is correct, since $h \in [g]$ implies $g^{-1}h \in S_0$, i.e., $A(g^{-1})x = x$, whence $A(h)x = A(g)x$. It follows from the strong continuity of the action and the openness of the canonical map $g \to [g]$ that f is continuous. It is also bijective : surjective thanks to the transitivity of A, and injective thanks to the implication $A(g)x = A(h)x \Rightarrow g^{-1}h \in S_0$. Since G/S_0 and X are compact, f is a homeomorphism. If now λ designates the left action of on G/S_0, then $f\lambda(h)[g] = f[hg] = A(hg)x = A(h)A(g)x = A(h)f[g]$, i.e., f transforms A into λ.

□

COROLLARY. *Suppose the compact group* G *acts effectively on the homogeneous space* X. *Then* X *is homeomorphic to* G *in such a way that the action of* G *on* X *is transformed into the left regular action of* G *on itself.*

□

The orbits of noncompact groups may have a more intricate topology ; in particular, they are not necessarily closed. A group action with dense orbits is called *topologically transitive.*

An action of a topological semigroup S *on a Hausdorff topological space* X *is, by definition, a strongly continuous homomorphism of* S *into the semigroup of all continuous maps* $X \to X$. If S is actually a topological group, then one reason (and the only one) the given semigroup action of S might not be a group action is that the image of the identity element of S is not the identity map of X (needless to say, however, every group action is also a semigroup action). The notion of an "action" depends on the category (topological groups or topological semigroups) for which it is defined.

The *orbits* of a semigroup action are defined as in the previous case ; however, from $x_2 \in O(x_1)$ it now follows only that $O(x_2) \subset O(x_1)$, and not necessarily the equality of the orbits $O(x_1)$ and $O(x_2)$: the orbits of two points may intersect without being identical.

Example. Let the additive semigroup $\mathbb{Z}_+$ of nonnegative integers act on itself by translations : $A(n)m = m + n$. The orbit of $m \in \mathbb{Z}_+$ is $\{k \mid k \in \mathbb{Z}_+, \ k \geq m\}$. Any two orbits intersect.

2°. Let G be a topological group, and let B be a Banach space. A *representation of* G *in* B is an action T of G on B such that the homeomorphism $T(g) : B \to B$ is linear, i.e., an automorphism of B, for every $g \in G$. Thus, a representation is a homomorphism $T : G \to \text{Aut } B$ which is strongly continuous in the sense that the vector-function $g \to T(g)x$ is continuous for every fixed $x \in B$.

Example 1. Let A be an invertible operator in B. Then $T(k) = A^k$, $k \in \mathbb{Z}$, is a representation of the group $\mathbb{Z}$. This in fact is the general form of the representations of $\mathbb{Z}$.

Example 2. Let A be an operator in B. Then $T(t) = e^{At}$, $t \in \mathbb{R}$, is a representation of the group $\mathbb{R}$. The family $\{e^{At}\}$ is called a *one-parameter group of operators*, and A is called its *generator* (A is also referred to as an *infinitesimal operator* in view of the formula $A = \frac{d}{dt}(e^{At})_{t=0}$). This terminology applies also to the original representation.

Example 3. Let Φ be a right invariant Banach space of functions on the group G such that the right translation $R(h)$ is strongly continuous in Φ. Then R is a representation, called the *right regular representation of* G *in* Φ. The notion of a *left regular representation* is defined in analogous manner.

Example 4. Every action A of the group G on a compact space X defines a representation A^* of G in $C(X)$, acting as $(A^*(g)\phi)(x) = \phi(A(g^{-1})x, \ \phi \in C(X)$.

A *representation of the topological semigroup* S *in the Banach space* B is an action T of S on B such that all the maps $T(s)$ are linear, i.e., endomorphisms of B. Thus, a representation of S in B is a homomorphism $T : S \to \text{End } B$ which is strongly continuous, meaning that the map $s \to T(s)x$ is continuous for every fixed $x \in B$.

If e is an identity element in S, then $T(e)$ is a projection (which commutes with all $T(s)$). A semigroup representation of a group S is a group representation only if $T(e) = E$.

Example 1. Let A be an arbitrary operator in B. Then $T(k) = A^k$, $k \in \mathbb{Z}_+$, is a a representation of the semigroup $\mathbb{Z}_+$. *This is the general form of the representations of* $\mathbb{Z}_+$ *satisfying* $T(e) = E$.

Example 2. Let S be a subsemigroup of the group G and let T be a representation of G. Then T can be restricted to S. In particular, for each $A \in L(B)$ one can consider the *one-parameter semigroup* $\{e^{At}\}$, $t \geq 0$, i.e., the representation $T(t) = e^{At}$ of the semigroup $\mathbb{R}_+$.

Example 3. Let $P : B \to B$ be a projection. The representation $T(s) \equiv P$ is called a *constant* representation, and in the cases $P = E$ and $P = 0$ a *unit* (or *identity*) representation and a *null* representation, respectively.

The *right regular representations* of a semigroup are defined in the same manner as for groups (left regular representation are not defined, since the rule $(L(s)\phi)(t) = \phi(st)$ would lead to $L(s_1 s_2) = L(s_2)L(s_1)$.

In the case of the semigroup $S = \text{End } B$ and of the group $G = \text{Aut } B$ the representation $T(s) \equiv s$, respectively $T(g) \equiv g$, is termed *trivial*.

From now on the definitions will be given, whenever possible, simultaneously for semigroups and groups. If the representation space B is finite dimensional (n-dimensional), the given representation is said to be *finite dimensional* (respectively, n-*dimensional*), and $n = \dim B < \infty$ is called its *dimension* or *degree*. If the representation space B is infinite dimensional, the representation is also called *infinite dimensional*. Representations for which the representation space is Hilbertian are called *Hilbert representations*. A representation is *isometric* (or, in the Hilbert case, *unitary*) if its image consists of isometric (respectively, unitary) operators. Finally, we say that a representation is *contractive* if its image consists of contractions. Every contractive group representation is isometric.

Example. Let G be a locally compact group and let dg be a right Haar measure on G. The right translation in $L_p(G)$ $(1 \leqslant p \leqslant \infty)$ is strongly continuous, and so it is a representation. Since the measure dg is invariant, this regular representation is isometric (and, in $L_2(G)$, unitary).

A representation T is said to be *bounded* if $\sup \|T(s)\| < \infty$. By the Banach-Steinhaus Theorem, this is equivalent to the boundedness of all orbits. Every representation of a compact semigroup is bounded.

LEMMA. *For every bounded representation T of a semigroup S in a Banach space B there exists an equivalent norm on B relative to which T is contractive.*

PROOF. Set $\|x\|' = \sup\limits_{s} \|T(s)x\|$, and $\|x\|_1 = \max(\|x\|, \|x\|')$ (in case that S has no identity element or $T(e) \neq E$). Then obviously $\|x\| \leqslant \|x\|_1 \leqslant c\|x\|$, where $c = \max(1, \sup\limits_{s} \|T(s\ \|)$. At the same time, $\|T(t)x\|' = \sup\limits_{s} \|T(st)x\| \leqslant \|x\|'$ and $\|T(t)x\| \leqslant \|x\|'$ for all $t \in S$. Consequently, $\|T(t)x\|_1 \leqslant \|x\|_1$, i.e., in the new norm T is contractive.

□

COROLLARY. *For every bounded representation T of a group G in a Banach space B there exists an equivalent norm on B relative to which T is isometric.*

□

Exercise. *Let T be a bounded representation of the semigroup S (group G). Then $\rho(T(s)) \leqslant 1$ for all s S (respectively, $\rho(T(g)) \equiv 1$ for all $g \in G$).*

Suppose given representations T_1 and T_2 in spaces B_1 and B_2, respectively. We say that T_2 is *equivalent* to T_1, and write $T_1 \sim T_2$, if there exists an isomorphism $F : B_1 \to B_2$ such that $T_2(s)F = FT_1(s)$ for all $s \in S$. Any morphism $F: B_1 \to B_2$ satisfying the last identity is said to *intertwine* T_2 and T_1. The classification of representations to within equivalence is one of the basic problems of representation theory.

As we already established, *every bounded representation of a semigroup (group) is equivalent to a contractive (respectively, isometric) representation.* However, under this equivalence a Hilbert representation becomes a Banach one.

THEOREM. *Let* S *be a right amenable semigroup and let* T *be a bounded Hilbert representation of* S *with the property that* $\inf \|T(s)x\| \geq \alpha\|x\|$ *(with* $\alpha > 0$*). Then* T *is equivalent to a unitary representation.*

PROOF. The argument uses the averaging method, which goes back to A. Hurwitz. Let H be the representation space of T. Put $\tau(s;x,y) = (T(s)x, T(s)y)$, $x,y \in H$, $s \in S$. The function τ is continuous and bounded in the variable s. Set $(x,y)_1 = \langle\tau(\cdot;x,y)\rangle$, where $\langle\cdot\rangle$ is an invariant mean on S. Then obviously $\alpha^2\|x\|^2 \leq (x,x)_1 \leq \beta^2\|x\|$, where $\beta = \sup_s \|T(s)\|$. Therefore, $(\cdot,\cdot)_1$ is a new inner product on H, and the norm that it defines is equivalent to the original norm $\|\cdot\|$. Relative to $(\cdot,\cdot)_1$, representation T is unitary: $(T(s)x, T(s)y)_1 = \langle(R(s)\tau)(\cdot;x,y)\rangle = \langle\tau(\cdot;x,y)\rangle = (x,y)_1$.

□

COROLLARY. *Every bounded Hilbert representation of an amenable group* G *is equivalent to a unitary representation.*

In fact, since $\|T(g^{-1})x\| \leq \beta\|x\|$, we have $\|T(g)x\| \geq \beta^{-1}\|x\|$, and the preceding theorem applies.

□

Corollary 1 encompasses a number of classical results which are listed below.

COROLLARY 2 (A. Hurwitz, 1897). *Every finite dimensional representation of a finite group* G *is equivalent to a unitary representation.*

□

COROLLARY 3 (H. Weyl, 1925). *Every Hilbert representation of a compact group* G *is equivalent to a unitary representation.*

□

[In Weyl's result G was a compact Lie group and the representation finite dimensional. The first condition was removed in works of A. Haar and A. Weil ; it actually was required only to guarantee the existence of an invariant measure. The second condition is trivially immaterial.]

COROLLARY 4 (A. Weil, 1940). *Every bounded finite dimensional representation* T *of an arbitrary group* G *is equivalent to a unitary representation.*

□

This assertion reduces to Corollary 3 since Im T is a compact group of operators with a trivial representation.

COROLLARY 5 (B. Sz.-Nagy, 1947). *Let* A *be an invertible operator in Hilbert space such that the powers* A^k, $k \in \mathbb{Z}$, *are bounded. Then* $A = F^{-1}UF$, *where* U *is unitary and* F *is an invertible operator.*

□

Here one uses the fact that isomorphic Hilbert spaces can be identified.

Exercise. *The operator* F *can be chosen self-adjoint and positive definite.*

COROLLARY 6 (M. G. Krein, 1964). *Every bounded Hilbert representation of an Abelian group* G *is equivalent to a unitary representation.* □

Let us pause to discuss an interesting problem, posed by S. Mazur and reproduced by S. Banach in the well-known annotations to his book [2], and subsequently by S. Ulam [48].

Consider the isometry group of a Banach space B. It acts on the unit sphere of B. If B is Hilbert, this action is transitive. In fact, if $\|x\| = \|y\| = 1$, then in the linear span L of the vectors x,y there exists a unitary operator U such that $Ux = y$, and one can extend U to a unitary operator in B by taking the identity operator in the orthogonal complement $L^{\perp}$.

MAZUR'S CONJECTURE. *Suppose the action of the isometry group of a separable Banach space* B *on the unit sphere of* B *is transitive. Then* B *is a Hilbert space.*

[In the nonseparable case a counterexample to this conjecture was offered by S. Rolewicz (1972). Incidentally, for some unclear reason Ulam ommited the separability requirement, though it was present in Mazur's formulation.]

This conjecture is still open in the general case, though in the finite dimensional case it is readily settled. In fact, the trivial representation of the isometry group of a finite dimensional space B with a norm $\|\cdot\|$, being bounded, is unitary with respect to some Hilbert norm $\|\cdot\|_1$. Pick an arbitrary point x_0 on the original unit sphere : $\|x_0\| = 1$. For every x with $\|x\| = 1$ there exists, by hypothesis, an isometry V of B such that $Vx_0 = x$. By the unitarity of the operator V, $\|x\|_1 = \|x_0\|_1\|x\|$. This identity can be extended by homogeneity from the unit sphere to the whole space, yielding $\|x\| = \text{const}\,\|x\|_1$ for all $x \in B$. Since $\|\cdot\|_1$ is a Hilbert norm, it follows that $\|\cdot\|$ is also a Hilbert norm. We thus proved the following

THEOREM (G. Auerbach, 1935). *If the isometry group of a finite dimensional normed space acts transitively on the unit sphere, then the space is Hilbert.*

□

It is interesting that *in the spaces* $L_p(0,1)$, $1 \leqslant p \leqslant \infty$, *the action of the isometry group is topologically transitive*

(S. Rolewicz, 1972). Therefore, Mazur's conjecture, if true, is a subtle fact.

Exercise. *In the space* ℓ_p, $1 \leq p \leq \infty$, *every isometry consists in a permutation of the terms of the sequence and their multiplication by numbers* ε_k *with* $|\varepsilon_k| = 1$.

In connection with Auerbach's Theorem we mention another open problem : *which closed subgroups of* $U(n)$ (*or* $O(n)$) *are isometry groups of an n-dimensional normed space ?* Some results concerning this problem were obtained by V. N. Kalyuzhnyi (1974, 1978).

3°. The importance of representations in comparison with other actions comes from the fact that, by definition, they enjoy the distinguishing property of linearity. The latter opens, in particular, the possibility of studying nonlinear objects (semigroups or groups) by means of linear "models". From this point of view it is interesting to single out the *faithful*, i.e., injective representations. However, such representations do not necessarily exist.

LEMMA. *Suppose the semigroup* S *possesses a faithful representation* T. *Then the continuous functions on* S *separate points.*

PROOF. Let B be the representation space of T. If $s_1 \neq s_2$, then $T(s_1) \neq T(s_2)$. Consequently, $T(s_1)x \neq T(s_2)x$ for some $x \in B$, and then there is a functional $f \in B^*$ such that $f(T(s_1)x) \neq f(T(s_2)x)$. The function $s \to f(T(s)x)$ is continuous on S and separates the points s_1 and s_2 (functions of this form are called *generalized matrix elements* of the representation T). □

COROLLARY. *There exists a semigroup that admits no faithful representations.*

In fact, every Hausdorff topological space can be transformed into a topological semigroup. At the same time, the space can be chosen so that the continuous functions do not separate its points.

As a matter of fact, one can even choose it so that all continuous functions are constant. Every representation of such a semigroup is constant.

□

In the class of bounded representations one can give a complete characterization of the semigroups that admit faithful representations.

THEOREM. *Let* S *be a semigroup. Then* S *admits a faithful bounded representation if and only if the right uniformly continuous functions on* S *separate points. If this condition is satisfied the representation can be chosen contractive.*

PROOF. NECESSITY. It is enough to convince ourselves that the generalized matrix elements are right uniformly continuous. But this follows from the estimate

$$|f(T(ut)x) - f(T(us)x)| \leq \beta\|f\| \, \|T(t)x - T(s)x\| .$$

SUFFICIENCY. Put $\tilde{S} = S$ if S has an identity element, and $\tilde{S} = S \cup \{e\}$ otherwise, adjoining e as a (topologically isolated) identity. Then S is a subsemigroup of $\tilde{S}$ and $\tilde{\phi} \in UCB_r(\tilde{S})$ if $\tilde{\phi}|S \in UCB_r(S)$. Hence, we may assume from the very beginning that S possesses an identity element e. In the Banach space $UCB_r(S)$ the right translation is a representation, since it is strongly continuous : $\|R(t)\phi - R(s)\phi\| = \sup_u \|\phi(ut) - \phi(us)\| \to 0$ as $t \to s$. This representation is obviously contractive. It remains to verify its faithfulness. Let $s_1 \neq s_2$. Then, by hypothesis, there is a function $\phi \in UCB_r(S)$ such that $\phi(s_1) \neq \phi(s_2)$, and then $(R(s_1)\phi)(e) \neq (R(s_2)\phi)(e)$, i.e., $R(s_1) \neq R(s_2)$.

□

COROLLARY 1. *Every locally compact semigroup admits a faithful contractive representation.*

□

This holds true, in particular, for compact and discrete semigroups. This means that there are no algebraic obstructions to the existence of faithful Banach representations : only the topology of the semigroup may be an obstruction.

COROLLARY 2. *Every group* G *admits a faithful isometric representation.*

In fact, by the Kakutani-Birkhoff Theorem, the algebra $UCB_r(G)$ separates the points of G.

□

It is interesting that there exist groups possessing no faithful unitary representations. As N. I. Nessonov (1983) showed, *in a Hilbert space the group of operators of the form* $E - A$, *where* A *runs through the set of all compact operators with no fixed points, admits no unitary representations apart from the identity representation.* In conjunction with Corollary 2 this suggests that the Banach-space framework is the natural one for the general theory of group representations. At the same time, *the regular representation in* $L_2(G)$ *is faithful and unitary for every locally compact group* G. Nonunitary Hilbert and, generally, Banach representations of locally compact groups deserve to be studied not so much as tools for learning about the groups themselves, but rather as objects arising independently. For example, the structure of the group $\mathbb{Z}$ is plain without resorting to representation theory. However, the set of all representations of $\mathbb{Z}$ in a Banach space B is in a one-to-one correspondence with Aut B. An analogous picture is valid for the semigroup $\mathbb{Z}_+$. In this sense the theory of representations of $\mathbb{Z}_+$ ($\mathbb{Z}$) is equivalent to the theory of operators (respectively, invertible operators), and in fact we already know that consideration of the powers of a given operator is useful in studying its properties. In the ensuing discussion we shall pursue deeper this point of view.

4°. The continuity property intervening in the definition of a representation can be strenghtened or relaxed. The representation T is said to be *uniformly* (or *norm*) *continuous* if it is continuous in the operator norm, i.e., it is a continuous homomorphism into Aut B , for a group, or into End B, for a semigroup. The simplest example is the trivial representation of Aut B or

End $\mathcal{B}$.

More interesting is the following

Example. *Let* $A \in \operatorname{End}\mathcal{B}$. *The representation* $t \to e^{At}$ *of the additve group* $\mathbb{R}$ *is uniformly continuous.*

Exercise. *The regular representation of* $\mathbb{R}$ *in* $L_2(\mathbb{R})$ *is not uniformly continuous.*

We consider next homomorphisms T which are continuous in the weak topology, i.e., such that all generalized matrix elements are continuous. In certain cases this suffices to guarantee that T is a representation.

THEOREM. *Let* G *be a locally compact group,* $\mathcal{B}$ *a Banach space, and* $T : G \to \operatorname{Aut}\mathcal{B}$ *a weakly continuous homomorphism. Then* T *is strongly continuous, i.e., a representation.*

[This result was published by K. de Leeuw and I. Glicksberg (1965), who refer to a private communication of H. Mirkil.]

PROOF. By hypothesis, all generalized matrix elements $f(T(g))$ are continuous, i.e., the vector-functions $g \to T(g)x$ are weakly continuous. Consequently, the integral

$$x_N(h) = \int_G \delta_N(g)T(hg^{-1})x\,dg \in \mathcal{B}$$

exists ; here δ_N denotes the indicator function of the arbitrary precompact neighborhood N of e. Obviously, $x_N(h) = T(h)x_N(e)$. We show that $x_N(h)$ converges strongly to $x_N(e)$ as $h \to e$. Consider the symmetric neighborhood $\Delta = NN^{-1}$ of e. Every function $f(T(g)x)$ is bounded on Δ. By the Banach-Steinhaus Theorem, $\sup_{g\in\Delta} \|T(g)\| = C < \infty$. Since

$$x_N(h) - x_N(e) = \int_\Delta [\delta_N(gh) - \delta_N(g)]T(g^{-1})x\,dg ,$$

it follows that

$$\|x_N(h) - x_N(e)\| \leqslant C \operatorname{mes} \Delta \sup_{g\in\Delta} |\delta_N(gh) - \delta_N(g)| ,$$

which gives the needed convergence. Now let B_0 denote the set of all vectors $y \in B$ such that $T(h)y \to y$ as $h \to e$. It suffices to show that $B_0 = B$. Notice that B_0 is a subspace of B: its linearity is plain, and its closedness follows from the boundedness of $\|T(h)\|$ in a neighborhood of e. Further, as we already saw, B_0 contains all elements $x_N(e)$. Since $f(x_N(e)) = \int_N \delta_N(g) f(T(g^{-1})x)\, dg$ and the function $f(T(g^{-1})x)$ is continuous, $f(x_N(e)) \to f(x)$, i.e., x is the weak limit of the net $\{x_N(e)\}$. This implies that $x \in B_0$, because every subspace of a Banach space is weakly closed. Since $x \in B$ was arbitrary, this proves that $B_0 = B$.

□

5°. The construction used in the preceding proof is widely applied in representation theory. It is connected with the so-called *functional* or *operational calculus* for a representation. We pause to discuss this important subject. Let T be a representation of the locally compact group G in the Banach space B. Consider the weight $\alpha_T(g) = \|T(g^{-1})\|$ (obviously, $\alpha_T(g_1g_2) \leqslant \alpha_T(g_1)\alpha_T(g_2)$). It defines the group algebra $L(G;\alpha_T) \equiv L(G;T)$. For each function $\phi \in L(G;T)$ there is defined (thanks to the continuity of the function $g \to T(g^{-1})x$ for every fixed $x \in B$) the operator

$$\boxed{\tilde{\phi} = \int \phi(g) T(g^{-1})\, dg \in L(B) ,}$$

called the *Fourier transform of the function* ϕ *under representation* T.

Example 1. Let $T(t) = e^{i\lambda t}$, $t \in \mathbb{R}$, be a one-dimensional representation of $\mathbb{R}$. Then $\alpha_T = \mathbb{1}$, the group algebra $L(\mathbb{R};T)$ is $L_1(\mathbb{R})$, and $\tilde{\phi} = \int_{-\infty}^{\infty} \phi(t) e^{-i\lambda t} dt$ is the classical Fourier transform of ϕ (the scalar $\tilde{\phi}$ is regarded as a multiplication operator in a one-dimensional vector space).

Example 2. Let T be the right (for definiteness) regular representation of the group G in $L_1(G)$. Then $\alpha_T = \mathbb{1}$, the group algebra $L(G;T)$ is $L_1(G)$, and

$$(\tilde{\phi}\psi)(h) = \int_G \phi(g)\psi(hg^{-1})\, dg ,$$

i.e., $\tilde{\phi}$ is the operator of convolution by ϕ : $\tilde{\phi}\psi = \phi * \psi$.

THEOREM. *The Fourier transformation* $\phi \to \tilde{\phi}$ *is a morphism of the group algebra* $L(G;T)$ *into the operator algebra* $L(B)$.

PROOF. The linearity of the Fourier transform is obvious. Its continuity follows from the bound $\|\tilde{\phi}\| \leqslant \|\phi\|$ (i.e., the Fourier transform is even contractive). Let us verify that the Fourier transform of a convolution is the product of the Fourier transforms of the factors, i.e., $(\phi * \psi)^{\sim} = \tilde{\phi}\tilde{\psi}$. Pick an arbitrary $x \in B$ and an arbitrary $f \in B^*$. Then

$$f((\phi * \psi)\, x) = \int \{\int \phi(h)\psi(gh^{-1})dh\}\, f(T(g^{-1})x)dg =$$

$$= \int \phi(h)dh \int \psi(gh^{-1})f(T(g^{-1})x)dg =$$

$$= \int \phi(h)dh \int \psi(g)f(T(h^{-1}g^{-1})x)dg =$$

$$= \int \phi(h)dh \int \psi(g)(T(h^{-1})*f)(T(g^{-1})x)dg =$$

$$= \int \phi(h)(T(h^{-1})*f)\tilde{\psi}x\, dh = \int \phi(h)f(T(h^{-1})\tilde{\psi}x)\, dh = f(\tilde{\phi}\tilde{\psi}x),$$

as claimed.

□

Exercise. $(R(h)\phi)^{\sim} = T(h)\tilde{\phi}$, *where* R *is the right regular representation of* G *in* $L(G;T)$, *i.e., the Fourier transform intertwines the given and the regular representations.*

We next consider the image of the group algebra under Fourier transform (referred to here as the *Fourier image*). It is a subalgebra $L_0(T) \subset L(B)$ which, generally speaking, is not closed even in the uniform topology. As a rule, $L_0(T)$ contains no operators of the form $T(g)$. However, all such operators belong to the strong closure of $L_0(T)$ (and to the uniform closure if the representation T is uniformly continuous). We may think of $T(g)$ as the Fourier transform of Dirac's δ-function supported at the point g^{-1}. The linear span $\mathrm{Lin}\, T \equiv \mathrm{Lin}(\mathrm{Im}\, T) = \mathrm{Lin}\, \{T(g)\}$ is the smallest subalgebra of $L(B)$ containing $T(g)$ for all $g \in G$.

2. DECOMPOSITION OF REPRESENTATIONS

The inner structure of a representation depends on the "supply" of its invariant subspaces. A representation T in a space B is said to be *reducible* if its image $\mathrm{Im}\, T$ is reducible, i.e., there exists a nontrivial (proper) subspace invariant under all operators $T(\cdot)$. Such a subspace L is said to be *invariant under representation* T, or T-*invariant* (this terminology applies also to $L = 0$ and $L = B$). In L there is defined the *subrepresentation* $T|L$. We remark that the reducibility property is preserved under passage to an equivalent representation.

For each vector $x \in B$ the smallest T-invariant subspace containing x is the closed linear span of its orbit $O(x)$. Hence, if T is irreducible, the linear span of the orbit $O(x)$ of every vector $x \neq 0$ is dense in B. If T is reducible, then there is an $x \neq 0$ such that the linear span of $O(x)$ is not dense in B (one can take any $x \neq 0$ in a T-invariant subspace $L \neq 0, B$).

Exercise 1. *Every irreducible representation of a finite semigroup is finite dimensional.*

Exercise 2. *Every irreducible finite dimensional representation of an Abelian semigroup is one-dimensional. Consequently, every irreducible representation of a finite Abelian semigroup is one-dimensional.*

One-dimensional representations are obviously always irreducible.

Example 1. *The one-dimensional representations of the group* $\mathbb{R}$ *have the form* $t \to e^{i\lambda t}$, *with* $\lambda \in \mathbb{C}$. *Such a representation is unitary if and only if* $\lambda \in \mathbb{R}$.

Example 2. *The one-dimensional representations of the group* $\mathbb{Z}$ *have the form* $n \to \zeta^n$, *with* $\zeta \in \mathbb{C}'$. *Such a representation is unitary if and only if* $\zeta \in \mathbb{T}$.

Example 3. *The one-dimensional representations of the group* $\mathbb{T}$ *have the form* $\zeta \to \zeta^n$, *with* $n \in \mathbb{Z}$. *They are all unitary.*

Irreducible representation are of special importance because under certain conditions they provide the "bricks" from which all other representations are built. We next discuss this aspect of representation theory in its simplest, essentially purely algebraic variant.

Suppose the space B of representation T decomposes into a direct sum of invariant subspaces : $B = L_1 \dot{+} \dots \dot{+} L_m$, $L_k \neq 0$, $k = 1,\dots,m$. Then we say that T *decomposes into the sum of subrepresentations* $T_k = T|L_k$, $k = 1,\dots,m$, and we write $T = T_1 \dot{+} \dots \dot{+} T_m$. A representation that admits at least one such decomposition (with $m > 1$) is said to be *decomposable*. Every decomposable representation is obviously reducible. The converse is however false.

Example. The representation $k \to \begin{pmatrix} 1 & k \\ 0 & 1 \end{pmatrix}$ of the additive semigroup $\mathbb{Z}_+$ in $\mathbb{C}^2$ is reducible but not decomposable.

A representation is said to be *completely reducible* if every invariant subspace has an invariant direct complement. Every completely reducible representation is clearly decomposable (or irreducible).

Example. *Every unitary representation* U *of a group* G *is completely reducible.* In fact, if L is an invariant subspace of U, then for every $g \in G$, L is invariant under the operator $U(g^{-1}) = U(g)^*$, and then $L^{\perp}$ is invariant under U(g).

The analogous statement for semigroups is false.

Example. In the Hilbert space $\ell_2 \equiv L_2(\mathbb{Z})$ consider the semigroup V^m, $m \geqslant 0$, generated by the unitary shift operator $V : (V\xi)_k = \xi_{k-1}$. The subspace L specified by the condition : $\xi_k = 0$ for all $k < 0$, is invariant under V, but not under V^{-1}. Hence, L has no invariant complement.

The properties of representations discussed above are conveniently formulated in terms of projections. To a decomposition of representation T into a sum of subrepresentations T_k, $1 \leqslant k \leqslant m$, there corresponds a resolution of identity $\{P_k\}_1^m$ (with P_k projections, $\sum_{k=1}^m P_k = E$, and $P_i P_k = 0$ if $i \neq k$) such that every P_k commutes with T, that is, with all operators $T(\cdot)$. This correspondence between decompositions of T and resolutions of identity which commute with T is bijective.

LEMMA. *The representation* T *is completely reducible if and only if for every* T-*invariant subspace* L *there is a projection* P *onto* L *which commutes with* T.

PROOF. Suppose T is completely reducible. Then we can take for P the projection onto L parallel to N, i.e., corresponding to the decomposition $B = L \dotplus N$, where N is a T-invariant complement of L. Conversely, if $PT(\cdot) = T(\cdot)P$, then Ker P is a T-invariant complement of L.

□

COROLLARY. *Every subrepresentation* T_1 *of a completely reducible representation* T *is completely reducible.*

In fact, let L be the representation space of T_1 and let

$M \subset L$ be a subspace invariant under T_1, and hence under T. If P is a projection onto M which commutes with T, then $P|L$ is a projection from L onto M which commutes with T_1.

□

A representation is called *semisimple* if it decomposes into a direct sum of irreducible subrepresentations.

THEOREM. *Every semisimple representation is completely reducible.*

This assertion is an easy consequence of the following

LEMMA. *Let* $T = T_1 \dot{+} \dots \dot{+} T_m$ *be a decomposition into a direct sum of irreducible representations,* $B = L_1 \dot{+} \dots \dot{+} L_m$ *the corresponding decomposition of the representation space* B *of* T, *and set* $L = L_2 \dot{+} \dots \dot{+} L_m$. *If* M *is a* T-*invariant subspace, then either* $M \subset L$ *or* $M + L = B$.

PROOF. Consider the associated resolution of identity $\{P_k\}_1^m$. The subspace $P_1M \subset L_1$ is T-invariant, since $T(\cdot)P_1y = P_1T(\cdot)y$ for all $y \in M$. But representation T_1 is irreducible, and so either $P_1M = 0$ or $P_1M = L_1$. In the first case it is plain that $M \subset L$. In the second case, pick any $x \in B$ and find an $y \in M$ such that $P_1y = P_1x$. Then $x - y \in L$. This shows that $M + L = B$.

□

PROOF OF THE THEOREM. Let $T = T_1 \dot{+} \dots \dot{+} T_m$ be a representation such that each T_k, $k = 1,\dots,m$, is irreducible. We proceed by induction on m. For $m = 1$ the needed assertion is trivial. Suppose it is true for $m-1$. Let M be a T-invariant subspace. Then in the first case of the alternative described by the lemma, M admits an invariant complement N in L. The subspace $L_1 \dot{+} N$ is then an invariant complement of M in B. In the second case $M \cap L$ admits an invariant complement N in L, and then N is an invariant complement of M in B.

□

One of the most important properties of semisimple representations is that their decomposition into irreducible representations is unique (to within permutations of the components and their replacement by equivalent representations).

UNIQUENESS THEOREM. *Suppose that the semisimple representation* T *admits two decompositions into a sum of irreducible subrepresentations* : $T = T_1 + \dots + T_m$ *and* $T = \hat{T}_1 + \dots + \hat{T}_\ell$. *Then* $m = \ell$, *and there is a permutation* $i_1,\dots,i_m$ *of* $1,\dots,m$ *such that the representations* $\hat{T}_{i_k}$ *and* T_k *are equivalent for all* $k = 1,\dots,m$.

PROOF. Assuming that $m \leq \ell$, we proceed by induction on m. For $m = 1$ the theorem is obvious. Suppose it is true for $m-1$. We use the preceding lemma and its notations, taking for M that of the representation (sub)spaces corresponding to $\hat{T}_1,\dots,\hat{T}_\ell$ which is not included in L. Suppose, for the sake of definiteness, that M is the representation space of $\hat{T}_1$. Since, by the irreducibility of $\hat{T}_1$, $M \cap L = 0$, we have $M \dot{+} L = B$. At the same time, $L_1 \dot{+} L = B$. This last decomposition yields a projection P with $\operatorname{Im} P = L_1$ and $\operatorname{Ker} P = L$. The isomorphism $P|M : M \to L$ intertwines $\hat{T}_1$ and T_1, since P commutes with T. Now let N denote the representation subspace for $\hat{T}_2 \dot{+} \dots \dot{+} \hat{T}_m$. The decomposition $M \dot{+} N = B$ yields a projection Q with $\operatorname{Im} Q = N$ and $\operatorname{Ker} Q = M$. The isomorphism $Q|L : L \to N$ intertwines $\hat{T}_2 \dot{+} \dots \dot{+} \hat{T}_m$ and $T_2 + \dots + T_\ell$. By the inductive hypothesis, we conclude that $m = \ell$ and that $\hat{T}_2,\dots,\mathbf{T}_m$ is a permutation of the collection $T_2,\dots,T_\ell$.

□

The Uniqueness Theorem permits us to speak about the irreducible components of a semisimple representation T and about their multiplicities in T. An irreducible representation $\hat{T}$ is called an *irreducible component* of the semisimple representation T if in the decomposition of T into a sum of irreducible subrepresentations there is a summand equivalent to $\hat{T}$. The number of such summands is called the *multiplicity with which* $\hat{T}$ *occurs in* T

or simply the *multiplicity of* $\hat{T}$ *in* T. If $T_1,\dots,T_r$ are all the irreducible components of the semisimple representation T and $\nu_1,\dots,\nu_r$ the corresponding multiplicities, we can write

$$T = \nu_1 T_1 + \dots + \nu_r T_r \; .$$

The summands $\nu_1 T_1,\dots,\nu_r T_r$ in this decomposition are called the *primary* or the *isotypical components* of the representation T. We have the following

THEOREM. *The resolution of identity associated with the decomposition of a semisimple representation* T *into primary components is unique.*

PROOF. Let $B = L_1 + \dots + L_r$, where the terms correspond to the primary components of T. Let M be a subspace such that $T|M \sim T_1$. Then $M \cap L_i = 0$ for all $i > 1$, whereas $M \cap L_1 \neq 0$: otherwise, B would contain the subspace $L_1 + M \neq L_1$ in which T_1 has multiplicity $\nu_1 + 1$. Thus, $M \subset L_1$ and the assertion of the theorem is immediate.

□

The tools used in the proof of the Uniqueness Theorem enable us to obtain a complete description of all subrepresentations of a semisimple representation.

THEOREM. *Let* T *be a semisimple representation and let* $T = T_1 \dotplus \dots \dotplus T_m$ *be a decomposition of* T *into irreducible components. Then every subrepresentation* $\hat{T}$ *of* T *is equivalent to a sum* $T_{i_1} \dotplus \dots \dotplus T_{i_r}$, *with* $i_1 < \dots < i_r$ *and* $r \geqslant 1$.

□

COROLLARY. *Every subrepresentation of a semisimple representation is semisimple.*

□

The notion of decomposition of a representation can be extended in one or other sense to the case of infinitely many components. We consider the simplest such situation. Let $\{P_\lambda\}$ be an arbitrary resolution of identity in the representation space B of T. If all projections P_λ commute with T, the subspaces $\operatorname{Im} P_\lambda$

are T-invariant. In this case we say that representation T decomposes into the *topological sum of the subrepresentations* $T_\lambda = T|\text{Im } P_\lambda$, or that T is *reduced* by the given resolution of identity $\{P_\lambda\}$.

We conclude this section by proving another general fact concerning irreducible representations.

THE SECOND SCHUR LEMMA. *Suppose the morphism* $F : B_1 \to B_2$ *intertwines two nonequivalent irreducible representations* T_1 *and* T_2. *Then* $F = 0$.

PROOF. Ker F is T_1-invariant. Since T_1 is irreducible, either Ker $F = B_1$, and then $F = 0$, or Ker $F = 0$, and then F is injective. In the second case, the subspace Im F is T_2-invariant and different from zero. Since T_2 is irreducible, Im $F = B_2$. Thus, F is bijective, and hence (by the theorem on the inverse operator) an isomorphism, which contradicts the nonequivalence of T_1 and T_2.

□

3. FINITE DIMENSIONAL REPRESENTATIONS

1°. In this section we give an exposition of the classical theory of finite dimensional representations, created by G. Frobenius, F. E. Molin, W. Burnside, and I. Schur at the end of the 19th and the beginning of the 20th century. Although the founders of this theory regarded the finite groups and their representations as its main subject, what they in fact mostly used was only the finite dimensionality or the semisimplicity of the representations. This permits us to extend a certain part of the classical theory to representations of an arbitrary topological semigroup S.

As we already know, the semisimplicity of a representation implies its complete reducibility. In the finite dimensional case these two properties are in fact equivalent.

THEOREM. *A finite dimensional representation is semisimple if and only if it is completely reducible.*

PROOF. The necessity of this condition has been already established (and for this it has not even been required that the representation be finite dimensional). Its sufficiency is obtained by induction on dimension : if $T = T_1 \dot{+} T_2$, then T_1, T_2 are completely reducible together with T, and hence semisimple, by the inductive hypothesis. Therefore, T is also semisimple.

□

COROLLARY. *Every finite dimensional unitary representation is semisimple.*

□

Remark. *Every finite dimensional unitary representation decomposes into an orthogonal sum of irreducible subrepresentations.*

For a finite group every finite dimensional representation is equivalent to a unitary one. Hence, *every finite dimensional representation of a finite group is semisimple.*

In the remaining part of this section we shall be concerned only with finite dimensional representations T of a semigroup S with identity element e (with $T(e) = E$).

Let $e_1, \ldots, e_n$ be a basis in the representation space of T. Then to each operator $T(s)$ there corresponds a matrix $(\tau_{ik}(s))$ $1 \leqslant i,k \leqslant n$ (in what follows we shall, as a rule, identify operators with their matrices). The elements $\tau_{ik}(\cdot)$ of this matrix, regarded as (obviously, continuous) functions on S, are called the *matrix elements* of the representation T (in the given basis).

THEOREM. *The family of matrix elements of an irreducible representation* T *is linearly independent.*

PROOF. Suppose $\sum_{i,k=1}^{n} \lambda_{ik}\tau_{ik}(s) \equiv 0$. Introducing the matrix of constants $\Lambda = (\lambda_{ik})_{i,k=1}^{n}$, we rewrite this identity as $\mathrm{tr}(\Lambda' T(s)) \equiv 0$ (where the prime stands for transposition and tr for trace). Passing to linear combinations of operators $T(s)$, we see that $tr(\Lambda' A) = 0$ for all $A \in$ Lin T. By Burnside's Theorem, Lin T is equal to the full algebra of linear operators. Consequently, $\Lambda = 0$.

□

The next problem we address to is the classification of finite dimensional representations to within equivalence. In the class of semisimple representations it admits a complete solution. We first consider a model (yet totally relevant) example.

Every representation T of the semigroup $\mathbb{Z}_+$ is uniquely specified by the operator $A = T(1)$, which in turn can be given arbitrarily. Since the irreducible subrepresentations of T are necessarily one-dimensional, for T to be semisimple it is necessary and sufficient that A be diagonalizable, i.e., that it admit a basis of eigenvectors. The problem of decomposing the given representation T into irreducible components becomes here the problem of decomposing with respect to the eigenvectors of the operator A, i.e., the basic problem of spectral theory of operators. Acccordingly, the classification of the representations of $\mathbb{Z}_+$ to within equivalence is equivalent to the classification of linear operators to within similarity. The solution to the latter is well known from linear algebra. In the general case it is formulated in terms of elementary divisors. In the class of diagonalizable operators a necessary and sufficient condition for similarity is that the operators have the same spectra and the same associated multiplicities. The equality of spectra with multiplicities accounted for is equivalent to the equality of the characteristic polynomials of the two operators in question. Passing to this invariant is indeed advisable, because the coefficients of the characteristic polynomial are rational (even polynomial) functions of the matrix elements of the operator, whereas in order to find the spectrum one has to actually solve the characteristic equation. However, in view of the well-known formulas of Newton, the coefficients of the characteristic equation can be written as polynomials in sums of powers of its roots (the eigenvalues) ; conversely, these sums are polynomials in the coefficients. As for the sums of powers of eigenvalues, they can be written as traces of powers of the operator A. Consider the function $\chi_T(k) = tr\, A^k$ on $\mathbb{Z}_+$. It is called the *character of the representation* T. We thus showed that two irreducible representations of the semigroup $\mathbb{Z}_+$ which have equal characters are equivalent. The converse is also true, and in fact obvious : similar operators have the same trace. Thus, the character of a semisimple representation of $\mathbb{Z}_+$ is a classifying invariant. A remarkable discovery of the classical representation theory is that this result carries over to arbitrary semigroups.

Thus, let T be an n-dimensional representation of the semigroup S (for the moment T is not necessarily semisimple). The function defined on S by the formula

$$\boxed{\chi_T(s) = \operatorname{tr} T(s)}$$

is called the *character* of T. The character is obviously invariant, meaning that the characters of equivalent representations coincide. The converse is false if the semisimplicity requirement is dropped. As a simple example we give the pair of nonequivalent representations of $\mathbb{Z}_+$ generated in two-dimensional space by the matrices $\begin{pmatrix}1 & 0\\ 0 & 1\end{pmatrix}$ and $\begin{pmatrix}1 & 1\\ 0 & 1\end{pmatrix}$: they have the same character, the constant function 2.

CLASSIFICATION THEOREM. *Two semisimple representations of the semigroup* S *are equivalent if and only if their characters coincide.*

PROOF. It remains to establish the sufficiency of this condition. En route we shall isolate, in the form of lemmas, a number of assertions of independent interest. In the space F(S) of all functions on S consider the linear span M_T of the matrix elements of representation T.

LEMMA 1. *The space* M_T *does not depend on the choice of a basis in the representation space of* T. *Also, it does not change on passing to an equivalent representation.*

PROOF. A function on S belongs to M_T if and only if it

has the form $\phi_\Lambda(s) = tr(\Lambda' T(s))$, where Λ $L(B)$. This decription does not depend on the choice of a basis and is invariant under substitution of T by an equivalent representation.

□

LEMMA 2. M_T *is biinvariant.*

PROOF. Under, say right translations :

$$\phi_\Lambda(st) = tr(\Lambda' T(st)) = tr(T(t)\Lambda' T(s)) = \phi_{\Lambda T'(t)}(s) .$$

□

LEMMA 3. *The character* χ_T *of* T *belongs to* M_T .

PROOF. $\chi_T(s) = \phi_E(s)$.

□

Now consider the linear span of the orbit of the character χ_T under the action of translations (left or right, it does not matter, since $\chi_T(st) = \chi_T(ts)$).

LEMMA 4. *Suppose representation* T *is irreducible. Then the linear span of the orbit of* χ_T *is equal to* M_T .

PROOF. Let $\phi_\Lambda(s)$ be an arbitrary function in M_T. By Burnside's Theorem, the operator Λ' is a linear combination of operators $T(t)$, $t \in S$: $\Lambda' = \sum_k \lambda_k T(t_k)$. Consequently, $\phi_\Lambda(s) = \sum_k \lambda_k \chi_T(st_k)$.

□

We remark that for an irreducible representation T a basis in M_T is provided by the matrix elements $\tau_{ik}(s)$, $1 \leqslant i,k \leqslant n$, thanks to their linear independence. Consequently, $\dim M_T = n^2$. By Lemma 2, a right regular representation R of S acts in M_T.

LEMMA 5. *Suppose* T *is irreducible. Then the representation* R *in* M_T *decomposes into a sum of* n *irreducible representations, each equivalent to* T.

PROOF. Let us write the action of R on the matrix elements of T :

$$(R(t)\tau_{ik})(s) = \tau_{ik}(st) = \sum_{j=1}^{n} \tau_{ij}(s)\tau_{jk}(t) ,$$

i.e.,

$$R(t)\tau_{ik} = \sum_{j=1}^{n} \tau_{jk}(t)\tau_{ij} , \quad 1 \leq i,k \leq n.$$

Therefore, for each fixed i the linear span L_i of the elements $\tau_{i1},\ldots,\tau_{in}$ is R-invariant and the matrix of the operator $R(t)|L_i$ in this basis is identical to the matrix of $T(t)$ in the basis $e_1,\ldots,e_n$. This means that the representation $R(t)|L_i$ is equivalent to T, and hence irreducible.

□

The last two lemmas admit the following

COROLLARY. *Let* T_1 *and* T_2 *be two irreducible representations such that* $\chi_{T_1} = \chi_{T_2}$. *Then* $T_1 \sim T_2$.

In fact, by Lemma 4, $M_{T_1} = M_{T_2}$. Denote this space by M. By Lemma 5, the regular representation of S in M admits two decompositions : $T_1 \dotplus \ldots \dotplus T_1 = T_2 \dotplus \ldots \dotplus T_2$ (in which the components are written to within equivalence). By the Uniqueness Theorem, $T_1 \sim T_2$.

□

The Classification Theorem is thus established for irreducible representations. To extend it to arbitrary semisimple representations we need the following lemma .

LEMMA 6. *Let* $T_1,\ldots,T_m$ *be pairwise nonequivalent irreducible representations of* S. *Then the family of matrix elements of* $T_1,\ldots,T_m$ *is linearly independent.*

PROOF. We proceed by induction on m. For $m = 1$ the lemma is already proven. Suppose it is valid for $m-1$. This means that the spaces $M_{T_1},\ldots,M_{T_{m-1}}$ are independent. Suppose now that the lemma is not valid for $T_1,\ldots,T_{m-1},T_m$. Then the intersection Π of M_{T_m} with the direct sum $\dot{\sum}_{k=1}^{m-1} M_{T_k}$ is different from zero.

The space Π is invariant under translations R. Since $\Pi \subset M_{T_m}$, all irreducible components of the representation $R|\Pi$ are equivalent (by Lemma 5) to T_m. On the other hand, it follows from the inclusion $\Pi \subset \dot{\sum}_{k=1}^{m-1} M_{T_k}$ that each of the irreducible components of $R|\Pi$ is equivalent to one of the representations $T_1,\dots,T_{m-1}$. This contradicts the Uniqueness Theorem, because $T_1,\dots,T_{m-1},T_m$ are pairwise nonequivalent.

□

COROLLARY 1. *Characters of pairwise nonequivalent irreducible representations are linearly independent.*

□

COROLLARY 2. *A finite semigroup* S *possesses only finitely many pairwise nonequivalent irreducible representations.*

□

We can now complete the proof of the Classification Theorem. Let T be a semisimple representation, $T_1,\dots,T_m$ the pairwise nonequivalent irreducible components of T, and $\nu_1,\dots,\nu_m$ their multiplicities. Then obviously

$$\chi_T = \sum_{k=1}^{m} \nu_k \chi_{T_k} .$$

By Corollary 1, the coefficients ν_k are uniquely determined by the character χ_T. Consequently, χ_T determines uniquely the representation T to within equivalence.

□

In the course of the proof we obtained valuable information on the structure of irreducible representations. As it turned out (Lemma 5), *every finite dimensional irreducible representation can be realized in a translation-invariant space of functions.* It is both interesting and useful to remark that these are the only possible realizations of the given representation in function spaces. More precisely, we have the following

THEOREM. *Let* Φ *be a right invariant finite dimensional space of functions on the semigroup* S *with identity* e, *such that the regular representation* R *of* S *in* Φ *is equivalent to an irreducible representation* T. *Then* $\Phi \subset M_T$.

PROOF. Let $\phi_1, \ldots, \phi_p$ be a basis in Φ, and let $\phi \in \Phi$. Then $\phi(s) = \sum_{k=1}^{p} \alpha_k \phi_k(s)$, whence $\phi(st) = \sum_{k=1}^{p} \alpha_k \sum_{i=1}^{p} \tau_{ik}(t)\phi_i(s)$. Consequently,

$$\phi(t) = \sum_{k=1}^{p} \alpha_k \sum_{i=1}^{p} \phi_i(e)\tau_{ik}(t),$$

i.e., $\phi \in M_T$.

□

From this theorem and Lemma 5 we obtain, in view of the Uniqueness Theorem, the following

COROLLARY. *If representation* $R|\Phi$ *is semisimple, then the multiplicity of any* n-*dimensional irreducible representation in* $R|\Phi$ *does not exceed* n.

□

2°. It is curious that there exist groups (not reduced to the identity) which admit no finite dimensional representation different from an identity representation. Such groups are necessarily infinite, as a regular representation of an arbitrary finite group is finite dimensional and faithful.

Example (I. Kaplansky, 1957). Consider the group of all bijections of the set $\mathbb{N} = \{1,2,3,\ldots\}$, i.e., of all permutations $k \to i_k$, where in the right-hand side each positive integer appears exactly once. A permutation is said to be *finite* if $i_k = k$ for all sufficiently large k. The finite permutations form a subgroup, in which, in turn, we single-out the subgroup E_∞ of even permutations : $E_\infty = \cup_{m=2}^{\infty} E_m$, where E_m designates the group of even permutations of degree m ; obviously, $E_m \subset E_{m+1}$. We claim that E_∞ *admits no finite dimensional representations other then identity ones.*

Let T be a finite dimensional representation of E_∞. Then T is faithful, since such are all its restrictions $T|E_m$ with $m \geq 5$, by virtue of the well-known theorem asserting that for

$m \geqslant 5$ the groups E_m are all simple (a proof of this theorem can be found, for example, in A. G. Kurosh's book [28]). Consequently, E_∞ is isomorphic to its image under T, i.e., to a group of linear operators in a finite dimensional vector space. It now suffices to exhibit an intrinsic property of E_∞ that is possessed by no linear group. In a linear group, given any chain $M_1 \subset M_2 \subset \ldots$ of subsets, the corresponding chain $M_1' \supset M_2' \supset \ldots$ of centralizers stabilizes, i.e., $M_k' = M_{k+1}'$ for all k larger than some k_0 (this is a consequence of finite dimensionality). On the other hand, in E_∞ such a stabilization does not occur for the chain of subsets E_m: in fact, the centralizer E_m' consists of all permutations that keep $1,\ldots,m$ fixed.

We say that a semigroup S *possesses sufficiently many finite dimensional representations* if the matrix elements of such representations separate the points of S. Later we will make acquaintance with important classes of semigroups with this property. At any rate, such is every finite semigroup, since it possesses faithful finite dimensional representations. It may nevertheless happen that there are not sufficiently many irreducible representations. For example, the semigroup with the multiplication law $st = s$ has only two irreducible representations - the identity and null representations, both one-dimensional.

In Kaplansky's example the group, while having the simplest possible topology - the discrete one - is algebraically complicated ; in any case, it is very far from being Abelian. In contradistinction to this one has the following

THEOREM. *Every locally compact (in particular, discrete or compact) Abelian group possesses sufficiently many one-dimensional unitary representations.*

□

This result is an important ingredient of the *duality theory* for groups (L. S. Pontryagin, 1934; E. R. van Kampen, 1935). It fails if the local compactness requirement is dropped. To see this one can take any connected 2-group (an example was given in Chap. 2, Sec. 1, 1°). Every one-dimensional representation χ of such a group is necessarily the identity representation, because $\chi^2 = 1$

and $\chi(e) = 1$.

The characters of all possible irreducible representations of the semigroup S will be referred to as *the characters of* S. The *dimension* (or *degree*) of the character χ is the dimension (degree) of the representation that generates χ. One-dimensional characters can be thus identified with the corresponding representations, i.e., with the continuous multiplicative homomorphisms $S \to \mathbb{C}$. Trivial examples are the *unit character* $\chi(s) \equiv 1$ and the *null character* $\chi(s) \equiv 0$. The one-dimensional character χ is said to be *unitary* if $|\chi(s)| \equiv 1$, and *semiunitary* if $|\chi(s)| \leq 1$ for all $s \in S$. If S possesses an identity e, then $\chi^2(e) = \chi(e)$, i.e., either $\chi(e) = 1$ or $\chi(e) = 0$. In the second case $\chi(s) = \chi(es) = \chi(e)\chi(s) = 0$ for all $s \in S$, i.e., χ is the null character.

<u>Exercise 1.</u> *Every bounded one-dimensional character is semiunitary. Thus, all one-dimensional characters of a compact semigroup are semiunitary.*

<u>Exercise 2.</u> *Every bounded one-dimensional character of a group is unitary. Thus, all one-dimensional characters of a compact group are unitary.*

The one-dimensional characters of any semigroup S form an Abelian semigroup, denoted $S^{\#}$, under pointwise multiplication. Its identity element is the unit character, its invertible elements are the characters that do not vanish, in particular, the unitary characters. The unitary characters of S form a group S^*. If G is a locally compact Abelian group, G^* is called the *dual group* (or the *character group*) of G. In this case G^* is canonically endowed with the *compact-open topology*, in which a neighborhood of an arbitrary $\chi_0 \in G^*$ is specified by a compact set $Q \subset G$ and an $\varepsilon > 0$ as $\{\chi \mid |\chi(g) - \chi_0(g)| < \varepsilon \ \forall g \in Q\}$. [The same topology can be introduced on $S^{\#}$ if the semigroup S is locally compact.] If G is compact this topology coincides with the topology induced by the inclusion $G^* \subset C(G)$.

Exercise 1. G^* *is a topological group.*

Exercise 2. $\mathbb{Z}^* = \mathbb{T}$, $\mathbb{T}^* = \mathbb{Z}$, *and* $\mathbb{R}^* = \mathbb{R}$ *(under canonical identifications).*

THEOREM. G *compact* $\Rightarrow$ G^* *discrete;* G *discrete* $\Rightarrow$ G^* *compact.*

PROOF. Suppose G is compact. Then in G^*, $N = \{\chi \mid |\chi(g)-1| < 1 \ \forall g \in G\}$ is a neighborhood of the identity. If $\chi \in N$, then $|\chi(g^k) - 1| < 1$ for all $g \in G$ and all $k \in \mathbb{Z}$. Put $\chi(g) = \lambda$. Then $|\lambda^k - 1| < 1$, $k = 1,2,3,\ldots$, whence $|\frac{1}{m}\sum_{k=1}^m \lambda^k - 1| < 1$, i.e., assuming $\lambda \neq 1$, $|\frac{1}{m}(\lambda^{m+1}-1)(\lambda-1)^{-1}-1| < 1$ for all $m = 1,2,3,\ldots$, which upon letting $m \to \infty$ leads to an absurd conclusion. Thus, $\lambda = 1$, i.e., $\chi(g) \equiv 1$, which means that $N = \{\mathbb{1}\}$, and hence that G^* is discrete.

Suppose now that G is discrete. Then the compact-open topology on G^* coincides with the pointwise-convergence topology, and so G^* becomes a subspace of $\mathbb{T}^G$. But $\mathbb{T}^G$ is compact by Tikhonov's Theorem, and G^* is closed in $\mathbb{T}^G$, as the subset defined by the system of equations $\chi(g_1g_2) - \chi(g_1)\chi(g_2) = 0$, $g_1, g_2 \in G$, and $\chi(e) = 1$, whose left-hand sides depend continuously on χ. Therefore, G^* is compact (and also a topological group, i.e., a compact group).

□

One can show that the group G^* is *locally compact* whenever G is a locally compact Abelian group. Furthermore, for each such G the rule $(ig)(\chi) = \chi(g)$ defines a *canonical homomorphism* $i : G \to G^{**}$. The culmination of duality theory is the following assertion.

THE DUALITY PRINCIPLE. *The canonical homomorphism* $i : G \to G^{**}$ *is a topological isomorphism.*

□

Exercise. *The duality principle is valid for* $\mathbb{Z}$, $\mathbb{T}$, *and* $\mathbb{R}$.

We shall not give here the proof of the general Duality Principle. It can be found in a number of books mentioned earlier ([5], [15], [19, Vol. I], [38], [49]). and also in the recently published monograph [34] especially devoted to this subject and the structure theorems intimately related to it.

For semigroup there is no duality theory which encompasses at least all finite semigroups. For example, *any finite semigroup* S *with one generator which possesses sufficiently many one-dimensional characters is a group.* This is in contrast to the fact that S admits (as every finite semigroup does) faithful finite dimensional representations.

LEMMA. *Suppose the semigroup* S *possesses sufficiently many one-dimensional characters. Then* S *is Abelian and separative, the latter meaning that the equalities* $s^2 = t^2 = st$ *imply* $s = t$.

The proof is an easy exercise.

□

THEOREM (E. Hewitt and H. S. Zuckerman, 1956). *Every separative discrete Abelian semigroup possesses sufficiently many one-dimensional characters.*

□

The proof is given in [7]. The theorem fails for locally compact semigroups. For instance, every semigroup consisting of idempotents is clearly separative. At the same time, it is readily seen that *if a semigroup consisting of idempotents possesses sufficiently many one-dimensional characters, then it is totally disconnected.* Consequently, in every semigroup possessing sufficiently many one-dimensional characters the subsemigroup of idempotents is totally disconnected. However, this condition, too, in conjunction with separativeness, fails to guarantee the existence of sufficiently many one-dimensional characters, even if we confine ourselves to the class of compact semigroups (E. I. Glazman, 1974).

To conclude this section, we introduce a number of terms used systematically.

Suppose the one-dimensional space spanned by a vector $x \neq 0$ is invariant under representation T of the semigroup S. Then

x is called a *weight vector* of T. In this case $T(s)x - \chi(s)x = 0$, where χ is a one-dimensional character (the continuity of χ follows from the formula $\chi(s) = f(T(s)x)/f(x)$, where $f \in B^*$, $f(x) \neq 0$). The character χ is called a *weight* of representation T. The invariant subspace $\{z \mid T(s)z - \chi(s)z = 0 \quad \forall s \in S\}$ associated with χ is called a *weight subspace*.

Exercise. *Let* $A \in L(B)$. *The weight vectors of the representation* $T(k) = A^k$ *of the semigroup* $\mathbb{Z}_+$ *are precisely the eigenvectors of* A, *and the associated weights have the form* $\chi(k) = \lambda^k$, *where* λ *are the corresponding eigenvalues.*

4. THE REPRESENTATION SPECTRUM OF AN ABELIAN SEMIGROUP

1°. The notion of spectrum of an operator admits a natural and useful generalization to representations. It is natural to define the *discrete spectrum* of representation T as the set $\operatorname{spec}_d T$ of its weights. A one-dimensional character χ of the semigroup S is called a *quasi-weight* of representation T if in the representation space B there is a net $\{x_\nu\}$ such that $\|x_\nu\| = 1$ for all ν and

$$T(s)x_\nu - \chi(s)x_\nu \to 0$$

for all $s \in S$ (in this case $\{x_\nu\}$ is termed a *quasi-weight net*). The set of all quasi-weights of representation T is called its *approximate spectrum* (or *a-spectrum*), denoted $\operatorname{spec}_a T$. The set of those quasi-weights for which one can find a sequence that serves as a quasi-weight net is called the *ω-approximate spectrum* or *ωa-spectrum*) of T, denoted $\operatorname{spec}_{\omega a} T$. We have the obvious inclusions

$$\operatorname{spec}_d T \subset \operatorname{spec}_{\omega a} T \subset \operatorname{spec}_a T \subset S^\# .$$

In the remaining part of this section we will consider only representations of Abelian semigroups.

Exercise 1. *Let* $A \in L(B)$ *and* $T(k) = A^k$, $k \in \mathbb{Z}_+$. *The formula* $\chi(k) = \lambda^k$ $(\lambda \in \text{spec}_a A)$ *establishes a bijective correspondence between* $\text{spec}_a A$ *and* $\text{spec}_a T$ $(= \text{spec}_{\omega a} T)$. *The same holds true for the case* $A \in \text{Aut}\, B$, $T(k) = A^k$, $k \in \mathbb{Z}$.

<u>Exercise 2.</u> *Let* $A \in L(B)$ *and* $T(t) = e^{At}$, $t \in \mathbb{R}$. *The formula* $\chi(t) = e^{\lambda t}$ $(\lambda \in \text{spec}\, A)$ *establishes a bijective correspondence between* $\text{spec}_a A$ *and* $\text{spec}_a T$ $(= \text{spec}_{\omega a} T)$.

<u>Remark.</u> The full *spectrum of representation* T can be defined as $\text{spec}\, T = \text{spec}_a T \cup \text{spec}_a T^*$. Despite the fact that $s \to T^*(s)$ is, generally speaking, merely a w^*-continuous homomorphism (rather than a representation), the notion of a quasi-weight can be defined for T^* in exactly the same manner as for T.

Let Σ be an Abelian semigroup of operators in B, endowed with the uniform operator topology. The definition of the a-*spectrum* $\text{spec}_a \Sigma$ (as well as of the other related notions) reduces to the previous one via the trivial representation.

<u>Exercise 1.</u> *Let* $\chi \in \text{spec}_a T$. *Then* $\chi(s) = \xi(T(s))$, *where* $\xi \in \text{spec}_a \text{Im}\, T$; *the quasi-weight* ξ *is uniquely determined.*

<u>Exercise 2.</u> *If representation* T *is uniformly continuous, the formula* $\chi(s) = \xi(T(s))$ *defines a bijection of* $\text{spec}_a \text{Im}\, T$ *onto* $\text{spec}_a T$.

One can define in analogous manner the a-spectrum and ωa-spectrum of any commutative family $\Phi \subset L(B)$ (the commutativity assumption is from now on in force).

<u>Exercise 1.</u> *Let* Φ *be a uniformly closed operator algebra. Every quasi-weight* ξ *of* Φ *is a multiplicative functional.*

<u>Exercise 2.</u> *Every quasi-weight* ξ *of the family* Φ *is uniformly continuous, extends to a quasi-weight of the smallest uniformly closed semigroup* $\Sigma \supset \Phi$, *and this extension is unique.*

<u>Exercise 3.</u> *Let* ξ *be a quasi-weight of the family* Φ. *Then* $\xi(A) \in spec_a A$ *for all* $A \in \Phi$. *Consequently,* $|\xi(A)| \leqslant \rho(A)$.

LEMMA. *The function* $\xi : \Phi \to \mathbb{C}$ *is a quasi-weight of the family* $\Phi \subset L(B)$ *if and only if for every* $\varepsilon > 0$ *and every finite subset* $F \subset \Phi$ *one can find a vector* $x_{F,\varepsilon} \in B$, $\|x_{F,\varepsilon}\| = 1$, *such that* $\|Ax_{F,\varepsilon} - \xi(A)x_{F,\varepsilon}\| < \varepsilon$ *for all* $A \in \Phi$.

The criterion for a character $\chi \in S^{\#}$ to be a quasi-weight of a representation T is formulated in exactly the same manner.

PROOF. Let ξ be a quasi-weight, and let $\{x_\nu\}$ be an associated net. Then for every $A \in F$ there is an index ν_A such that $\|Ax_\nu - \xi(A)x_\nu\| < \varepsilon$ for all $\nu > \nu_A$. Now pick any μ such that $\mu > \nu_A$ for all $A \in F$ and put $x_{F,\varepsilon} = x_\mu$.

Conversely, $\{x_{F,\varepsilon}\}$ turns into a quasi-weight net upon ordering the family of all finite subsets $F \subset \Phi$ by inclusion, the half-line $\varepsilon > 0$ in decreasing order, and finally the set of pairs (F,ε) componentwise. In fact, for every $A \in \Phi$ and every $\varepsilon > 0$ the conditions $\{A\} \subset F$ and $0 < \eta < \varepsilon$ imply $\|Ax_{F,\eta} - \xi(A)x_{F,\eta}\| < \varepsilon$.

□

We topologize $spec_a\Phi$ endowing the space of all functions $\Phi \to \mathbb{C}$ with the topology of pointwise convergence.

LEMMA. *The a-spectrum of the family* Φ *is compact.*

PROOF. First, notice that $spec_a\Phi$ is contained in the set of all functions $\theta : \Phi \to \mathbb{C}$ that satisfy the inequality $|\theta(A)| \leqslant \rho(A)$ for all $A \in \Phi$. The latter is compact by Tikhonov's Theorem. Secondly, $spec_a\Phi$ is closed. In fact, if the function $\theta \in spec_a\Phi$, then there exist an $\varepsilon > 0$ and a finite subset $F \subset \Phi$ such that given any $x \in B$, $\|x\| = 1$, one can find an $A_x \in F$ with the property that $\|A_x x - \theta(A_x)x\| \geqslant \varepsilon$. But then $\|A_x x - \phi(A_x)x\| \geqslant \frac{\varepsilon}{2}$ for every ϕ in the neighborhood $N_{F,\varepsilon} = \{\phi \mid |\phi(A) - \theta(A)| < \varepsilon/2 \ \forall A \in F\}$. Therefore, $N_{F,\varepsilon} \cap spec_a\Phi = \emptyset$, i.e., the complement of $\text{spec}_a \Phi$ is open.

□

COROLLARY. *The a-spectrum of any representation* T *is closed in the pointwise-convergence topology on* S *(and hence in every stronger topology, in particular, in the compact-open one). If* T *is uniformly continuous, then* $\text{spec}_a T$ *is compact.*

The same holds true for the ωa-spectrum.

□

We now address the following crucial question : is the spectrum always nonempty ? An affirmative answer has been given by the author (1971) for the ωa-spectrum of a uniformly separable family. V. I. Lomonosov (1973) proved that the a-spectrum is not empty in the general case. Z. Slodkowski and W. Zelasko (1974), and Y. Domar and L. Lindhal (1975) proposed other approaches in the Banach algebra framework (they also introduced other notions of a spectrum and subjected them to a comparative analysis). The proof of the nonemptiness of the ωa-spectrum relies on a geometric construction which processes the essential spectrum of an arbitrarily given operator into a discrete spectrum, thereby transforming the original problem into one of the same kind as the algebraic problem of finding a common eigenvector for a commuting family of matrices. However, the underlying space remains infinite dimensional and for this reason we do not end up with a purely algebraic problem.

Let $m(B)$ denote the Banach space of all bounded sequences $X = \{x_k\}_1^\infty \subset B$, equipped with the standard norm $\|X\| = \sup_k \|x_k\|$. In $m(B)$ we single-out the subspace $c_0(B) = \{X \mid \lim_{k\to\infty} x_k = 0\}$, and then consider the quotient space $\hat{m}(B) = m(B)/c_0(B)$. The image of the element $X \in m(B)$ in $\hat{m}(B)$ will be denoted by $\hat{X}$.

LEMMA 1. $\|\hat{X}\| = \overline{\lim_{k\to\infty}} \|x_k\|$.

PROOF. Setting $X_k = \{\underbrace{0,\ldots,0}_{k},x_{k+1},\ldots\}$, we have $X - X_k = \{x_1,\ldots,x_k,0,\ldots\} \in c_0(B)$. Consequently, $\hat{X} = \hat{X}_k$ and $\|\hat{X}\| = \|\hat{X}_k\| \leqslant \sup_{q>k} \|x_q\|$. If $\|\hat{X}\| \geqslant \|X\| - \varepsilon$, then a fortiori $\|\hat{X}\| \geqslant$

$\geq \sup_{q>k} \|x_q\| - \varepsilon$. The needed equality follows upon letting $k \to \infty$ and $\varepsilon \to 0$.

□

Now let $A \in L(B)$. We extend A to $m(B)$, setting $\tilde{A}X = \{Ax_k\}_1^\infty$.

LEMMA 2. *The mapping* $A \to \tilde{A}$ *is a Banach algebra isometry* $L(B) \to L(m(B))$.

[This isometry is clearly not onto.]

PROOF. That the mapping $A \to \tilde{A}$ enjoys the required algebraic properties is obvious. Next, we have

$$\|\tilde{A}X\| = \sup_k \|Ax_k\| \leq \|A\| \sup_k \|x_k\| = \|A\|\|X\|,$$

whence $\|\tilde{A}\| \leq \|A\|$. On the other hand, $\|\tilde{A}\| \geq \|A\|$ since the $\tilde{A}$-invariant subspace $m_{const}(B)$ of constant sequences is canonically isometric to B, and this isometry intertwines A and $\tilde{A}|m_{cons}(B)$.

□

The subspace $c_0(B)$ is $\tilde{A}$-invariant. Therefore, the quotient operator $\hat{A}$ in $\hat{m}(B)$ is correctly defined as $\hat{A}\hat{X} = (\tilde{A}X)^\wedge$.

LEMMA 3. *The mapping* $A \to \hat{A}$ *is a Banach algebra isometry* $L(B) \to L(\hat{m}(B))$.

PROOF. Again, it suffices to verify that $\|\hat{A}\| = \|A\|$. By Lemma 1,

$$\|\hat{A}\hat{X}\| = \overline{\lim_{k\to\infty}} \|Ax_k\| \leq \|A\| \overline{\lim_{k\to\infty}} \|x_k\| = \|A\| \|\hat{X}\|,$$

whence $\|\hat{A}\| \leq \|A\|$. If now $x_k = x$ for all k, then $\|\hat{A}\hat{X}\| = \|Ax\|$, which gives $\|\hat{A}\| \geq \|A\|$.

□

The fact that one can use algebraic arguments is a consequence of the following result.

LEMMA 4. $\operatorname{spec}_a A = \operatorname{spec}_d \hat{A}$.

PROOF. The relation $Ax_k - \lambda x_k \to 0$ is equivalent to $\tilde{A}X - \lambda X$

$\in c_0(B)$, i.e., to the equality $\hat{A}\hat{X} - \lambda\hat{X} = 0$. Also, $\overline{\lim}_{k\to\infty} \|x_k\| = 1$ means that $\|\hat{X}\| = 1$.

□

BASIC LEMMA. *Suppose the operators* $A_1,\ldots,A_m$ *pairwise commute. Then every collection* $\{\lambda_1,\ldots,\lambda_{m-1}\} \subset \mathrm{spec}_{\omega a}(\{A_1,\ldots,A_{m-1}\})$ *extends to a collection* $\{\lambda_1,\ldots,\lambda_{m-1},\lambda_m\} \subset \mathrm{spec}_{\omega a}(\{A_1,\ldots,A_{m-1},A_m\})$.

PROOF. Since A_m commutes with A_k, it follows that $\hat{A}_m$ commutes with $\hat{A}_k$, $k \leq m-1$. Consequently, the subspace $L_{m-1} = \{\hat{X} \mid \hat{A}_k\hat{X} = \lambda_k \hat{X}_k,\ k \leq m-1\}$ is $\hat{A}_m$-invariant. Hence, there exists a $\lambda_m \in \mathrm{spec}_a(\hat{A}_m | L_{m-1})$. Let $\{\hat{X}^{(k)}\}$ be a sequence such that $\|\hat{A}_m\hat{X}^{(k)} - \lambda_m\hat{X}^{(k)}\| < \frac{1}{k}$, $\|\hat{X}^{(k)}\| = 1$, $k = 1,2,3,\ldots$, and $\hat{A}_j\hat{X}^{(k)} - \lambda_j\hat{X}^{(k)} = 0$ for all j, $1 \leq j \leq m-1$. Let $X^{(k)} = \{x_n^{(k)}\}_{n=1}^{\infty}$. Then

$$\overline{\lim_{n\to\infty}} \|A_m x_n^{(k)} - \lambda_m x_n^{(k)}\| < \frac{1}{k}, \qquad \overline{\lim_{n\to\infty}} \|x_n^{(k)}\| = 1,$$

and

$$\lim_{n\to\infty} \|A_j x_n^{(k)} - \lambda_j x_n^{(k)}\| = 0 .$$

Now let n_k be a positive integer such that $\|A_j x_n^{(k)} - \lambda_j x_n^{(k)}\| < \frac{1}{k}$, $1 \leq j \leq m$, for all $n > n_k$. Pick $\nu_k > n_k$ such that $\frac{1}{2} < \|x_{\nu_k}^{(k)}\| < \frac{3}{2}$, and set $y_k = x_{\nu_k}^{(k)}$. Then

$$\|A_j y_k - \lambda_j y_k\| < \frac{1}{k}, \qquad 1 \leq j \leq m,$$

and

$$\frac{1}{2} < \|y_k\| < \frac{3}{2} .$$

Therefore, $Y = \{y_k\} \in m(B)$, $\hat{Y} \neq 0$, and $\hat{A}_j\hat{Y} - \lambda_j\hat{Y} = 0$, $1 \leq j \leq m$.

□

COROLLARY. *The* ω*-spectrum of any finite family* Φ *is nonempty.*

In fact, when the size $|\Phi| = 1$ we already know this from the spectral theory of operators, and the Basic Lemma permits an induction with respect to $|\Phi|$.

□

We are now ready to prove the following

THEOREM. *1) The* ωa*-spectrum of any separable commutative family* $\Phi \subset L(B)$ *is nonempty.*

2) The a*-spectrum of any commutative family* $\Phi \subset L(B)$ *is nonempty.*

PROOF. 1) Suppose Φ is countable : $\Phi = \{A_k\}_1^\infty$. Then by the Basic Lemma there exists a sequence of numbers $\Lambda = \{\lambda_k\}_1^\infty$ such that $\{\lambda_1,\dots,\lambda_m\} \subset \mathrm{spec}_{\omega a}(\{A_1,\dots,A_m\})$ for all m. Using the diagonal process it is now readily established that $\Lambda \subset \mathrm{spec}_{\omega a}\Phi$, and then $\Lambda \subset \mathrm{spec}_{\omega a}\overline{\Phi}$, where $\overline{\Phi}$ denotes the uniform closure of Φ.

2) Consider the set of functions $\xi \in \mathrm{spec}_a\Psi$, with $\Psi \subset \Phi$, ordered in the natural manner : $\xi_1 \leqslant \xi_2$ if $\Psi_1 \subset \Psi_2$ and $\xi_2|\Psi_1 = \xi_1$. If $\{\xi_\nu\}$ is a linearly ordered subset in this set we denote $\Psi = \cup_\nu \Psi_\nu$, and then, choosing for each $A \in \Psi$ an arbitrary ν such that $A \in \Psi_\nu$, we put $\xi(A) = \xi_\nu(A)$. It is clear that this construction does not depend on the choice of ν and yields a function ξ such that $\xi \geqslant \xi_\nu$ for all ν. We claim that $\xi \in \mathrm{spec}_a\Psi$. In fact, if $F \subset \Psi$ and $|F| < \infty$, then there is a ν such that $F \subset \Psi_\nu$. Then $\xi|F = \xi_\nu$ by construction. Let $\varepsilon > 0$. There is a vector $x \in B$, $\|x\| = 1$, such that $\|Ax - \xi_\nu(A)x\| < \varepsilon$, i.e., $\|Ax - \xi(A)x\| < \varepsilon$ for all $A \in F$. Using Zorn's Lemma, we conclude that there is a maximal function $\xi_0 : \Psi_0 \to \mathbb{C}$, $\xi_0 \in \mathrm{spec}_a\Psi_0$. It remains to show that $\Psi_0 = \Phi$. Let $A \in \Phi \setminus \Psi_0$ and $\widetilde{\Psi}_0 = \Psi_0 \cup \{A\}$. For any finite set $F \subset \Psi_0$ we put $\widetilde{F} = F \cup \{A\}$. Now consider all possible extensions of the function $\xi_0|F$ to elements of $\mathrm{spec}_a\widetilde{F}$. The existence of such extensions is guaranteed by the Basic Lemma. Their values at the point A form a compact set Q_F, included in the compact set $\mathrm{spec}_a A$, which does not depend on F. The family $\{Q_F\}$ is centered, since obviously

$Q_{F_1} \cap \ldots \cap Q_{F_n} \supset Q_{F_1 \cap \ldots \cap F_n}$. Consequently, there is a $\lambda_0 \in \cap_F Q_F$. Define a function $\tilde{\xi}_0 : \tilde{\Psi}_0 \to \mathbb{C}$ by $\tilde{\xi}_0|F_0 = \xi_0$ and $\tilde{\xi}_0(A) = \tilde{\lambda}_0$. We show that $\tilde{\xi}_0 \in \mathrm{spec}_a \Psi_0$, thereby contradicting the maximality of ξ_0. Let $F \subset \Psi_0$, $|F| < \infty$. Then there exists a $\xi \in \mathrm{spec}_a \tilde{F}$ such that $\xi|F = \xi_0|F$ and $\xi(A) = \lambda_0$, i.e., $\xi = \tilde{\xi}_0|\tilde{F}$. The vector $x_{\tilde{F},\varepsilon}$ that corresponds to the function ξ works also for $\tilde{\xi}_0$.

□

Remark 1. It follows from the proof that *the canonical projections* $\mathrm{spec}_a \Phi \to \mathrm{spec}_a A$ $(A \in \Phi)$ *are surjective*. Moreover, *every quasi-weight of an arbitrary subfamily* $\Psi \subset \Phi$ *can be extended to a quasi-weight of* Φ.

Remark 2. *In the separable case the a-spectrum and ωa-spectrum coincide.*

COROLLARY. *The a-spectrum of any uniformly continuous representation* T *of an Abelian semigroup* S *is nonempty. If, in addition,* S *is separable, then the ωa-spectrum of* T *is nonempty.*

PROOF. As we already know, if $\xi \in \mathrm{spec}_a \mathrm{Im}\, T$ and $\chi(s) = \xi(T(s))$, $s \in S$, then $\chi \in \mathrm{spec}_a T$. The situation for the ωa-spectrum is similar.

□

The uniform continuity and separability requirements are essential.

Example. Consider the previously encountered connected 2-group G consisting of the measurable subsets of the segment $I = [0,1]$ with the symmetric difference operation $\oplus$ and the metric $\mathrm{mes}(M \oplus N)$. We introduce $\{-1,1\}$-valued indicator functions of sets, putting $\tau_M(t) = -1$ if $t \in M$ and $\tau_M(t) = 1$ if $t \notin M$. Then $\tau_{M\oplus N} = \tau_M \tau_N$. Now consider the multiplication operators $T(M)$: $T(M)\phi = \tau_M \phi$ in $L_2(I)$. The mapping $M \to T(M)$ is a (unitary) representation of G, but is not uniformly continuous: indeed, $\|T(M) - T(N)\| = 2$ whenever $M \neq N$. *The a-spectrum of* T *is empty* : G possesses only the unit character, which cannot be a quasi-weight of T since $T(I) = -E$. *The a-spectrum of the family* $\operatorname{Im} T$ *is of course nonempty.* It is a 2-group algebraically isomorphic to G, but, in contrast to G, discrete (in the uniform topology). The characters of the group $\operatorname{Im} T$ assume the values ± 1 and separate its points. Let ξ be one of these characters. We show that ξ is a quasi-weight of the family $\operatorname{Im} T$. Pick an arbitrary finite collection of sets $M_1, \ldots, M_p$ and an arbitrary $\varepsilon > 0$. Set $N_k = M_k$ if $\xi(M_k) = 1$ and $N_k = I \setminus M_k$ if $\xi(M_k) = -1$. Set $N = N_1 \cup \ldots \cup N_p$. Let ϕ be a function equal to 1 outside N and, in case $\operatorname{mes} N = 1$, also on a set of measure ε^2, and equal to zero at the remaining points. Then $\|T(M_k)\phi - \xi(M_k)\phi\| \leqslant 2\varepsilon$, $1 \leqslant k \leqslant p$, i.e., ξ is indeed a quasi-weight of $\operatorname{Im} T$. It is not hard to show that the ωa-spectrum of the family $\operatorname{Im} T$ is empty.

2°. Let S be an Abelian semigroup with identity e, and let T be a representation of S in a space B with $T(e) = E$. The uniform-closed linear span $\overline{\operatorname{Lin} T}$ of the family $\operatorname{Im} T$ is a commutative Banach algebra of operators. The following result is readily established.

LEMMA. *Every character* $\chi \in \operatorname{spec}_a T$ *can be written in the form* $\chi(s) = \xi(T(s))$, *where* ξ *is a uniquely determined multiplicative functional on* $\overline{\operatorname{Im} T}$.

□

This exhibits a natural mapping $\operatorname{spec}_a T \to M(\overline{\operatorname{Lin} T})$. It is obviously injective and, as is readily verified, continuous. Consequently, $\operatorname{spec}_a T$ is homeomorphic to a compact subset of $M(\overline{\operatorname{Lin} T})$, and it is convenient to identify $\operatorname{spec}_a T$ with the latter. On the other hand, $\operatorname{spec}_a T \subset S^{\#}$, by definition. We would therefore like to regard the maximal ideal space $M(\overline{\operatorname{Lin} T})$ as a subset of $S^{\#}$, i.e., to have that : *1) the formula* $\chi(s) = \xi(T(s))$ *gives a character of* S *for every* $\xi \in M(\overline{\operatorname{Lin} T})$, and *2) the mapping* $\xi \to \chi$ *is continuous* (notice that it is always injective). *To guarantee that 1) holds it suffices to assume that* T *is uniformly continuous ; 2) is satisfied if* $S^{\#}$ *is endowed with the pointwise-convergence topology.* Under these assumptions $M(\overline{\text{Lin } T})$ can be identified with a subset of $S^{\#}$, called the *δ-spectrum* of the representation T and denoted $\text{spec}_\delta T$. Obviously,

$$\text{spec}_a T \subset \text{spec}_\delta T \subset S^{\#} .$$

Example. Let A be an operator in a Hilbert space H which maps H isometrically onto a subspace $H_1 \neq H$. Consider the representation $T(k) = A^k$ of the semigroup $\mathbb{Z}_+$. $\mathbb{Z}_+^{\#}$ is topologically isomorphic to the multiplicative semigroup $\mathbb{C}$. Here $\text{spec}_\delta T$ and $\text{spec}_a T$ can be identified with the unit disk $\mathbb{D}$ and the unit circle $\mathbb{T}$, respectively.

THEOREM. *Let* T *be a uniformly continuous representation of the Abelian semigroup* S. *Suppose that every character in the δ-spectrum of* T *is semiunitary. Then every unitary character* $\chi \in \text{spec}_\delta T$ *belongs to* $\text{spec}_a T$.

PROOF. Let $\chi \in \text{spec}_\delta T$ be unitary. Pick an arbitrary finite collection $\{s_1,\dots,s_m\} \subset S$ and consider the operator $A = \frac{1}{m}\sum_{k=1}^{m} \overline{\chi(s_k)}T(s_k)$. It belongs to the algebra $\overline{\text{Lin } T}$. Since T is uniformly continuous, $\chi(s) = \xi(T(s))$, where ξ is a multiplicative functional on $\overline{\text{Lin } T}$. It follows from the unitarity of χ that $\xi(A) = 1$. At the same time, $|\eta(A)| \leqslant 1$ for every multiplicative functional η, thanks to the semiunitarity of the character $\eta(T(s))$. Thus, the spectral radius of A in $\overline{\text{Lin } T}$ is equal to 1, and $\lambda = 1$ belongs to the spectrum of A. Consequently, the spectral radius of the operator A (i.e., of the element $A \in L(B)$) does not exceed 1. However, A - E is not an invertible operator: otherwise, we would have $(A - E)^{-1} = \lim_{\lambda \downarrow 1} (A - \lambda E)^{-1} \in \overline{\text{Lin } T}$, contradicting the noninvertibility of A - E in $\overline{\text{Lin } T}$. Thus, $1 \in \text{spec } A$ and $\rho(A) = 1$. Consequently, $1 \in \text{spec}_a A$. It follows that there exists a $\zeta \in \text{spec}_a \overline{\text{Lin } T}$ such that $\zeta(A) = 1$. Put

$\zeta_k = \zeta(T(s_k))$, $1 \leq k \leq m$. It follows from the definition of A that $\sum_{k=1}^{m} \overline{\chi(s_k)}\zeta_k = m$. Since $|\chi(s_k)| = 1$ and $|\zeta_k| \leq 1$ (the latter because of the semiunitarity of the character $\psi(s) = \zeta(T(s)))$, we have that $\zeta_k = \chi(s_k)$, $1 \leq k \leq m$. We see that in every neighborhood of the character χ (in the pointwise-convergence topology) there is a character $\psi \in \mathrm{spec}_a T$. Since the a-spectrum is closed, this implies that $\chi \in \mathrm{spec}_a T$.

□

COROLLARY 1. *Let* T *be a uniformly continuous representation such that any character in the* δ*-spectrum of* T *is unitary. Then* $\mathrm{spec}_\delta T = \mathrm{spec}_a T$.

□

We say that representation T is of *null exponential type* if $\lim_{n\to\infty} n^{-1}\ell n\|T(s^n)\| = 0$ for all $s \in S$ (this definition does not presume that T is uniformly continuous). By Gelfand's formula, this is the same as saying that all operators T(s) have spectral radius 1. If the uniformly continuous representation T is of null exponential type, then every character $\chi \in \mathrm{spec}_\delta T$ is semiunitary : in fact, $\chi(s) \in \mathrm{spec}_a T(s)$, and so $|\chi(s)| \leq \rho(T(s)) = 1$ [the same holds true for all $\chi \in \mathrm{spec}_a T$, even if we drop the uniform-continuity assumption (which is formally used to ensure that $M(\overline{\mathrm{Lin}\ T}) \subset S^{\#}$)]. We thus have

COROLLARY 2. *Let* T *be a uniformly continuous representation of null exponential type. Then every unitary character in* $\mathrm{spec}_\delta T$ *belongs to* $\mathrm{spec}_a T$.

□

COROLLARY 3 (Lyubich-Matsaev-Fel'dman, 1973). *The* δ*-spectrum and* a*-spectrum of any uniformly continuous representation of null exponential type of an Abelian group coincide.*

In fact, every semiunitary group character is unitary:

□

Next, guided by the example of an individual operator A (i.e., of the representation $k \to A^k$), we introduce the notion of of a *spectral subspace* of a representation T. This is defined as any T-invariant subspace L which contains all T-invariant subspaces M with the property that $\mathrm{spec}(T|M) \subset \mathrm{spec}(T|L)$, where spec designates one of the type of representation spectra, specified beforehand. A compact set $Q \subset S^{\#}$ is called a *spectral compact set* if there exists a spectral subspace L such that $\mathrm{spec}(T|L) = Q$ [$S^{\#}$ is endowed with any of the natural topologies ; it is assumed that all spectra considered are compact in $S^{\#}$]. Finally, T is called a *representation with separable spectrum* if the collection of all spectral compact sets of T is a basis for the topology of the spectrum ; a sufficient condition for this to be the case is that every compact set which is the closure of its interior be spectral. The separability of spectrum property is an analogue of complete reducibility, naturally fitted for nonunitary representations with nondiscrete spectrum. At any rate, it is clear that *any representation whose spectrum is separable and does not reduce to a single point is reducible*. Chapter 5 will be devoted to representations with separable spectrum.

CHAPTER 4

REPRESENTATIONS OF COMPACT SEMIGROUPS

1. HARMONIC ANALYSIS ON COMPACT GROUPS

1°. The classical prototype (and a particular case) of harmonic analysis on a compact group is the theory of odinary Fourier series. We survey it briefly from this point of view.

The unit circle $\mathbb{T}$ is a compact Abelian group. It has a regular representation in the group algebra $L_1(\mathbb{T})$. The one-dimensional subspaces $E_n = \mathrm{Lin}(e^{int})$, $n \in \mathbb{Z}$, are invariant, and so $\mathbb{T}$ has irreducible representations T_n in E_n. These are pairwise nonequivalent. To the family of representations $\{T_n\}$ there corresponds the resolution of identity $\{P_n\}$, where P_n is the projection onto E_n, which sends each function ϕ into its n-th Fourier harmonic $P_n\phi$. Thus, with each function $\phi \in L_1(\mathbb{T})$ one can associate the Fourier series $\phi \sim \sum_{n=-\infty}^{\infty} P_n\phi$, which is precisely the classical Fourier series of ϕ.

Every function $\phi \in L_1(\mathbb{T})$ is the limit in L_1-norm of a sequence of linear combinations of exponentials e^{int} (and in fact it suffices to include in these combinations only those exponentials which appear with nonzero coefficients in the Fourier series of ϕ). If $\phi \in C(\mathbb{T})$, then it is a uniform limit of linear combinations of exponentials (Weierstrass's Theorem). If $\phi \in L_2(\mathbb{T})$

the Fourier series of ϕ converges in $L_2(\mathbb{T})$ (the Riesz-Fischer Theorem). The resolution of identity $\{P_n\}$ is orthogonal, and so

$$\|\phi\|^2 = \sum_{n=-\infty}^{\infty} \|P_n\phi\|^2$$ (the *Parseval equality*).

Exercise. *Every irreducible representation of* $\mathbb{T}$ *is equivalent to one of the representations* T_n.

On passing to an arbitrary compact group this picture is basically preserved, but its details become more intricate, and its justification requires more elaborate tools. The construction of harmonic analysis on an arbitrary compact group was carried out in 1927 by H. Weyl in collaboration with F. Peter (except for the existence of an invariant measure, established later). This important branch of mathematics is presently known as the *Peter-Weyl theory*. We should emphasize that a key role in this theory is played by the procedure of averaging with respect to an invariant measure, on the basis of which I. Schur has reconstructed, at the beginning of this century, the theory of representations of finite groups.

Throughout this section G will denote a compact group and dg the normalized Haar measure on G.

KEY LEMMA. *Every unitary representation* U *of the group* G *possesses a finite dimensional invariant subspace different from zero.*

PROOF. Let H be the representation space and let A be a compact self-adjoint operator in H. For $x,y \in H$ we put

$$a(x,y) = \int (AU(g)x, U(g)y)\,dg .$$

The Hermitian bilinear functional a is bounded : $|a(x,y)| \leqslant \|A\|\,\|x\|\|y\|$. Consequently, $a(x,y) = (\tilde{A}x,y)$, where $\tilde{A}$ is a bounded ($\|\tilde{A}\| \leqslant \|A\|$) self-adjoint operator. We prove that $\tilde{A}$ is compact. To this end it suffices to show that $\tilde{A}$ takes any sequence $\{x_k\}_1^\infty \subset H$ which converges weakly to zero into a sequence which converges strongly to zero. We have

$$\|\tilde{A}x_k\|^2 = \int (AU(g)x_k, U(g)\tilde{A}x_k)\,dg \ .$$

The sequence of functions under the integral sign is bounded by the constant $M^2\|A\|^2$, where $M = \sup_k \|x_k\|$. Also, it converges to zero for every $g \in G$ because $U(g)x_k \overset{w}{\to} 0$, and so $AU(g)x_k \to 0$ thanks to the compactness of the operator A. By Lebesgue's theorem on passing to the limit under the integral sign, $\|\tilde{A}x_k\| \to 0$. Suppose now that $A \geqslant 0$, $A \neq 0$. Then also $\tilde{A} \geqslant 0$, $\tilde{A} \neq 0$. In fact,

$$(\tilde{A}x, x) = \int (AU(g)x, U(g)x)\,dg \geqslant 0 \ ,$$

and if $(Ax_0, x_0) > 0$, then $(\tilde{A}x_0, x_0) > 0$ too, since for $x = x_0$ the integrand is continuous and nonnegative on G, and positive at $g = e$. By the Spectral Theorem, $\tilde{A}$ has an eigenvalue $\lambda > 0$, and the corresponding eigensubspace L is finite dimensional. But $\tilde{A}$ commutes with all the operators $U(h)$ by its construction. Consequently, L is invariant under U.

□

COROLLARY 1. *Every irreducible unitary representation of* G *is finite dimensional.*

COROLLARY 2. *Every irreducible unitary representation of an Abelian group* G *is one-dimensional.*

□

We now turn to the important task of decomposing unitary representations of a group G into irreducible representations.

THEOREM (A. Gurevich, 1943). *Let* U *be a unitary representation of the group* G *in a Hilbert space* H. *Then* H *admits an orthogonal sum decomposition*

$$H = \overline{\bigoplus_\nu H^{(\nu)}} \qquad (1)$$

in which every $H^{(\nu)}$ *is invariant and the subrepresentations* $U|H^{(\nu)}$ *are irreducible.*

[The family $H^{(\nu)}$ is, generally speaking, infinite : its cardinality is equal to the dimension of H, interpreted as the cardinality of any orthonormal basis of H.]

The original formulation of this theorem included the requirement that G be first-countable, which is superfluous.

PROOF. Let us agree to call *partial decomposition* of the representation U any nonempty family $\Delta = \{H^{(\mu)}\}$ of pairwise orthogonal U-invariant subspaces of H such that the restrictions $U|H^{(\mu)}$ are irreducible. The Key Lemma guarantees the existence of a partial decomposition. The set of all partial decompositions is ordered by inclusion, and is inductive, i.e., every totally ordered subset $\{\Delta_\alpha\}$ has a majorant, namely, $\Delta = \bigcup_\alpha \Delta_\alpha$. By Zorn's Lemma, there exists a maximal partial decomposition $\Delta_{max} = \{H^{(\nu)}\}$. We claim that equality (1) holds for Δ_{max}. Suppose this in not the case. Then the orthogonal complement $[\bigoplus_\nu H^{(\nu)}]^\perp$ is different from zero and U-invariant, and by the Key Lemma it contains an invariant subspace H_0 such that $U|H_0$ is irreducible. But then $\Delta_{max} \cup \{H_0\}$ is a partial decomposition of U bigger that Δ_{max} : contradiction.

□

COROLLARY. *If G is Abelian, then in H there exists an orthonormal basis consisting of weight vectors of representation U.*

□

<u>Exercise.</u> *Let V be an irreducible subrepresentation of the representation U, and let $L \subset H$ denote the corresponding invariant subspace. Then $L \subset \overline{\oplus H^{(\nu)}}$, where the sum is taken over all indices ν such that $U|H^{(\nu)} \sim V$.*

In Sec. 2 we will obtain a Banach space analogue of Gurevich's Theorem, which will play an important role in the sequel.

2°. Let us apply the decomposition theorem to the right regular representation R of the group G in $L_2(G)$. It yields the decomposition

$$L_2(G) = \bigoplus_\nu E^{(\nu)} \tag{1_2}$$

where $E^{(\nu)}$ are R-invariant and the subrepresentations $R|E^{(\nu)}$ are irreducible. We claim that *every function in* $E^{(\nu)}$ *is continuous, possibly after altering its values on a set of measure zero.* It suffices to show this for functions $\varepsilon_1,\ldots,\varepsilon_m$ which form an ortnonormal basis in $E^{(\nu)}$. Let $\tau_{ik}(g)$ denote the matrix elements of representation $R|E^{(\nu)}$. They are continuous functions on G. Moreover,

$$\varepsilon_k(hg) = \sum_{i=1}^{m} \tau_{ik}(g)\varepsilon_i(h) \;, \quad 1 \leqslant k \leqslant m \;, \tag{2}$$

for every g and a.e. h. We let Γ denote the set of all pairs (g,h) for which (2) holds. Then Γ is a measurable subset of $G \times G$ and every "vertical" section of Γ obtained by fixing $g \in G$ has measure 1. By Fubini's Theorem, mes $\Gamma = 1$. Consequently, every "horizontal" section of Γ obtained by fixing h has measure 1. Therefore, there is an $h \in G$ such that (2) holds for a.e. g, and so

$$\varepsilon_k(g) = \sum_{i=1}^{m} \tau_{ik}(h^{-1}g)\varepsilon_i(h) \;, \quad 1 \leqslant k \leqslant m \;,$$

for a.e. $g \in G$. Our claim now follows from the fact that the right-hand sides of these equalities are continuous. Henceforth we shall assume that each subspace $E^{(\nu)}$ consists of continuous functions.

Now consider the set $\{V_\lambda\}$ of all pairwise-nonequivalent irreducible unitary representations of the group G. In the Euclidean space E_λ of representation V_λ choose an orthonormal basis $\varepsilon_\lambda^{(1)},\ldots,\varepsilon_\lambda^{(n_\lambda)}$, where $n_\lambda = \dim E_\lambda$. Let $\{\tau_{\lambda,ik}(g)\}_{i,k=1}^{n_\lambda}$ denote the corresponding matrix elements. It turns out that they satisfy the *orthogonality relations*

$$\boxed{(\tau_{\lambda,ik},\tau_{\mu,j\ell}) = \frac{1}{n_\lambda}\,\delta_{\lambda\mu}\delta_{ij}\delta_{k\ell}\ .}$$

To prove them we again resort to averaging over G. Let A be an arbitrary linear operator from E_μ into E_λ. Set

$$F = \int V_\lambda(g) A V_\mu^*(g)\,dg.$$

Then the operator F intertwines V_μ and V_λ. By the Second Schur Lemma, $F = 0$ whenever $\mu \neq \lambda$. If $\mu = \lambda$, then F is scalar by the First Schur Lemma, and $\operatorname{tr} F = \operatorname{tr} A$. Thus,

$$\int V_\lambda(g) A V_\mu^*(g)\,dg = \begin{cases} 0, & \mu \neq \lambda, \\ \left(\frac{1}{n_\lambda} \operatorname{tr} A\right) \operatorname{id}_{E_\lambda}, & \mu = \lambda. \end{cases}$$

Putting $Ax = (x, e_\mu^{(\ell)}) e_\lambda^{(k)}$ for $x \in E_\mu$, we have

$$\int V_\lambda(g) A V_\mu^*(g) e_\mu^{(j)}\,dg = \sum_i (\tau_{\lambda,ik}, \tau_{\mu,j\ell}) e_\lambda^{(i)},$$

which in conjunction with the previous relations yields the needed orthogonality of the matrix elements.

Returning to the decomposition (1_2), we let M_λ denote the linear span of the matrix elements $\tau_{\lambda,ik}$ $(1 \leqslant i,k \leqslant n_\lambda)$. By the general theory of finite dimensional representations, every subspace $E^{(\nu)}$ with the property that $R|E^{(\nu)} \sim V_\lambda$ is contained in M_λ. Since M_μ is orthogonal to M_λ for $\mu \neq \lambda$, those $E^{(\nu)}$ for which $R|E^{(\nu)}$ is not equivalent to V_λ are orthogonal to M_λ. Therefore, M_λ is the orthogonal sum of all subspaces $E^{(\nu)}$ such that $R|E^{(\nu)} \sim V_\lambda$ (and there are n_λ such $E^{(\nu)}$'s). Finally,

$$L_2(G) = \overline{\bigoplus_\lambda M_\lambda},$$

and so we proved the following result.

THE PETER-WEYL THEOREM. *The matrix elements* $\tau_{\lambda,ik}$ *form an orthogonal basis in* $L_2(G)$.

□

Generally speaking, $\tau_{\lambda,ik}$ are not normalized : $\|\tau_{\lambda,ik}\|^2 = n_\lambda$. Since $L_2(G)$ is dense in $L_1(G)$ and the L_2-norm is stronger than the L_1-norm, we obtain the following

COROLLARY. *The system of matrix elements* $\tau_{\lambda,ik}$ *is dense in*

$L_1(G)$.

□

Thus, $L_1(G) = \overline{\sum_\lambda M_\lambda}$. It turns out that this is a topological direct sum, i.e., it is associated with a certain resolution of identity. We will establish this fact below, after we derive a number of important formulas of the theory of characters.

3°. Let χ_λ denote the character of the unitary representation V_λ. An immediate consequence of the orthogonality of the matrix elements is that the charaters χ_λ are orthogonal :

$$\boxed{(\chi_\lambda, \chi_\mu) = \delta_{\lambda\mu} .}$$

LEMMA. *The following relations hold :*

$$\boxed{\chi_\lambda * \chi_\mu = \frac{1}{n_\lambda} \delta_{\lambda\mu} \chi_\mu .}$$

PROOF. We have :

$$(\chi_\lambda * \chi_\mu)(h) = \int \chi_\lambda(g)\chi_\mu(hg^{-1})dg =$$

$$= \sum_{i,k=1}^{n_\lambda, n_\mu} \int \tau_{\lambda,ii}(g)\tau_{\mu,kk}(hg^{-1})dg .$$

Since

$$V_\mu(hg^{-1}) = V_\mu(h)V_\mu^*(g),$$

it follows that

$$\tau_{\mu,kk}(hg^{-1}) = \sum_{j=1}^{n_\mu} \tau_{\mu,kj}(h)\overline{\tau_{\mu,kj}(g)} ,$$

and hence that

$$\int \tau_{\lambda,ii}(g)\tau_{\mu,kk}(hg^{-1})dg = \sum_{j=1}^{n_\mu} \tau_{\mu,kj}(h) \int \tau_{\lambda,ii}(g)\overline{\tau_{\mu,kj}(g)}dg =$$

$$= \sum_{j=1}^{n_\mu} \tau_{\mu,kj}(h)(\tau_{\lambda,ii}, \tau_{\mu,kj}) =$$

$$= \frac{\delta_{\lambda\mu}\delta_{ik}}{n_\lambda} \sum_{j=1}^{n_\lambda} \delta_{ij}\tau_{\mu,kj}(h) = \frac{1}{n_\lambda} \delta_{\lambda\mu}\delta_{ik} \tau_{\mu,ki}(h) .$$

Consequently,

$$(\chi_\lambda * \chi_\mu)(h) = \frac{1}{n_\lambda} \delta_{\lambda\mu} \sum_{i,k=1}^{n_\lambda, n_\mu} \delta_{ik}\tau_{\mu,ki}(h) = \frac{1}{n_\lambda} \delta_{\lambda\mu}\chi_\mu(h) ,$$

as asserted.

□

Set $\pi_\lambda = n_\lambda \chi_\lambda$. We have the following

COROLLARY. $\boxed{\pi_\lambda * \pi_\mu = \delta_{\lambda\mu}\pi_\mu .}$

□

Now in $L_1(G)$ consider the family of operators Π_λ : $\Pi_\lambda \phi = \pi_\lambda * \phi$ (= $\phi * \pi_\lambda$, since the characters belong to the center of the group algebra). Obviously,

$$\boxed{\Pi_\lambda \Pi_\mu = \delta_{\lambda\mu}\Pi_\mu ,}$$

i.e., $\{\Pi_\lambda\}$ is an algebraically-orthogonal family of projections.

THEOREM. *The image of* Π_λ *is the linear span* M_λ *of the matrix elements* $\tau_{\lambda,ik}$ $(1 \le i,k \le n_\lambda)$. *The family* $\{\Pi_\lambda\}$ *is a resolution of identity in* $L_1(G)$ *which reduces the regular representation (both right and left - recall that the subspaces* M_λ *are biinvariant).*

PROOF. By definition,

$$(\Pi_\lambda \phi)(h) = \int \phi(g)\pi_\lambda(hg^{-1})dg =$$

$$= n_\lambda \sum_{i,k=1}^{n_\lambda} \tau_{\lambda,ik}(h) \int \phi(g)\overline{\tau_{\lambda,ik}(g)}\, dg .$$

This clearly implies that $\operatorname{Im} \Pi_\lambda \subset M_\lambda$, and since $\Pi_\lambda \tau_{\lambda,j\ell} = \tau_{\lambda,j\ell}$ (by the orthogonality relations), $\operatorname{Im} \Pi_\lambda = M_\lambda$.

The completeness and algebraic orthogonality of the family $\{\Pi_\lambda\}$ have been thus established. It remains to verify that $\{\Pi_\lambda\}$ is a total family. Suppose $\pi_\lambda * \phi = 0$ for all λ. Then $\pi_\lambda *(\phi * \delta_Q) = 0$ for every compact subset $Q \subset G$, where δ_Q denotes the indicator function of Q. The function $\phi * \delta_Q$ is continuous and hence belongs to $L_2(G)$. But $\Pi_\lambda | L_2(G)$ *is the orthogonal projection onto* M_λ, because

$$\Pi_\lambda \psi = n_\lambda \sum_{i,k=1}^{n_\lambda} (\psi, \tau_{\lambda,ik}) \tau_{\lambda,ik}$$

for all $\psi \in L_2(G)$. Therefore, if $\psi \in L_2(G)$ and $\Pi_\lambda \psi = 0$ for all λ, then $\psi = 0$ by the Peter-Weyl Theorem. Consequently, $\phi * \delta_Q = 0$, i.e., $\int_Q \phi(g)dg = 0$ for any compact Q, which in turn gives $\phi = 0$.

□

COROLLARY 1. *The system of matrix elements* $\tau_{\lambda,ik}$ *is complete in* $C(G)$.

□

Thus, *every continuous function on the group* G *can be arbitrarily well uniformly approximated by linear combinations of matrix elements* $\tau_{\lambda,ik}$. This generalization of the Weierstrass Theorem is known as the *Peter-Weyl C-Theorem*, to distinguish it from the L_2-theorem discussed earlier. To prove it, it suffices to remark that if a complex measure σ on G annihilates all matrix elements, then it also annihilates all convolutions $\pi_\lambda * \phi$ with $\phi \in C(G)$. But then the "convolution" $(\phi * \sigma)(g) = \int \phi(hg^{-1})d\sigma \in C(G)$ is annihilated by all projections Π_λ, and hence it is equal to zero. Since $\phi \in C(G)$ is arbitrary, this gives $\sigma = 0$.

Remark. One can give a constructive proof of the Peter-Weyl C-Theorem using sliding means.

Exercise. *One has the topological direct sum decomposition* $C(G) = \overline{\Sigma_\lambda M_\lambda}$. *An analogous decomposition holds for every space* $L_p(G)$ *with* $1 < p < \infty$.

COROLLARY 2. *The group G possesses sufficiently many finite dimensional (unitary) representations.*

In fact, by Corollary 1 the matrix elements $\tau_{\lambda,ik}$ separate the points of G.

□

<u>Exercise.</u> *We say that the topological group G has no small subgroups if there exists a neighborhood of the identity element e which contains no subgroups different from {e}. A compact group G admits a faithful finite dimensional representation if and only if it has no small subgroups.* This implies, in particular, that *every compact Lie group admits a faithful finite dimensional representation.*

Now we correspond to each function $\phi \in L_1(G)$ its *Fourier series*

$$\phi \sim \sum_\lambda \Pi_\lambda \phi = \sum_\lambda n_\lambda \sum_{i,k=1}^{n_\lambda} \tau_{\lambda,ik} \int \phi(g)\overline{\tau_{\lambda,ik}(g)}\, dg .$$

Since the system of projections $\{\Pi_\lambda\}$ is total, ϕ *is uniquely determined by its Fourier series.* If $\phi \in L_2(G)$, then the Fourier series of ϕ converges to ϕ in $L_2(G)$. This generalization of the *Riesz-Fischer Theorem* is a straightforward consequence of the Peter-Weyl L_2-Theorem. It in turn yields the *generalized Parseval equality*

$$\|\phi\|^2 = \sum_\lambda n_\lambda \sum_{i,k=1}^{n_\lambda} \left| \int \phi(g)\overline{\tau_{\lambda,ik}(g)}\, dg \right|^2 .$$

4°. Let $Z_1(G)$ denote the center of the group algebra $L_1(G)$. It contains all characters χ_λ. Since $Z_1(G)$ is a subalgebra, the projections Π_λ map $Z_1(G)$ into itself. Hence, $\Pi_\lambda\phi \in M_\lambda \cap Z_1(G)$ for every $\phi \in Z_1(G)$. Write the function $(\Pi_\lambda\phi)(g)$ in the form $\mathrm{tr}(AV_\lambda(g))$, where A is a constant operator in the representation space of V_λ. Then the condition that it be a central function takes the form :

$$\operatorname{tr}(AV_\lambda(gh)) = \operatorname{tr}(AV_\lambda(hg)), \quad \text{i.e.,} \quad \operatorname{tr}([A,V_\lambda(g)]V_\lambda(h)) = 0 ,$$

where [,] denotes the commutator of operators. It now follows from the irreducibility of representation V_λ and Burnside's Theorem that $[A,V_\lambda(g)] = 0$. This in turn implies, by the First Schur Lemma, that $A = cE$. Consequently, $\Pi_\lambda \phi = c\chi_\lambda$, which in view of the form of $\Pi_\lambda \phi$ gives

$$\int \phi(g)\overline{\tau_{\lambda,ik}(g)}\, dg = cn_\lambda^{-1}\delta_{ik} .$$

Setting here $i = k$ and summing over i we get $c = \int \phi(g)\overline{\chi_\lambda(g)}dg$. Thus, *the Fourier series of a central function* ϕ *has the form*

$$\boxed{\phi \sim \sum_\lambda \chi_\lambda \int \phi(g)\overline{\chi_\lambda(g)}\, dg ,}$$

i.e., it coincides with the formal expansion of ϕ with respect to the orthonormal system of characters $\{\chi_\lambda\}$.

THEOREM. *The characters* χ_λ *form an orthonormal basis in the Hilbert space* $Z_2(G) = Z_1(G) \cap L_2(G)$.

PROOF. For $\phi \in Z_2(G)$ the Fourier series has the form $\phi \sim \sum_\lambda (\phi,\chi_\lambda)\chi_\lambda$. If all the Fourier coefficients (ϕ,χ_λ) vanish, then $\phi = 0$. Hence, the system $\{\chi_\lambda\}$ is complete in $Z_2(G)$.
□

COROLLARY 1. *The system of characters* $\{\chi_\lambda\}$ *is complete in* $Z_1(G)$.

In fact, $Z_2(G)$ is dense in $Z_1(G)$ in the L_1-metric.
□

Now in $C(G)$ consider the subspace $Z(G) = Z_1(G) \cap C(G)$ of central functions.

COROLLARY 2. *The system of characters* $\{\chi_\lambda\}$ *is complete in*

$Z(G)$.

PROOF. Suppose that the finite complex measure σ on G annihilates all the characters χ_λ. Then $\phi * \sigma = 0$ for all $\phi \in Z(G)$, as in the proof of the Peter-Weyl C-Theorem. But in the present case $\phi * \sigma$ is continuous. The equality $(\phi * \sigma)(e) = 0$ means that σ annihilates ϕ.

□

The value $\chi_\lambda(g)$ of the character χ_λ at g depends only on the conjugacy class of g. Hence, characters are continuous functions on the compact space $\tilde{G}$ of conjugacy classes. The space $C(\tilde{G})$ can be identified with $Z(\tilde{G})$. Corollary 2 is now restated as follows.

COROLLARY 2'. *The system of characters* $\{\chi_\lambda\}$ *is complete in* $C(\tilde{G})$. *As a consequence, the characters* χ_λ *separate the conjugacy classes of the group* G. *On an Abelian group* G *the characters* χ_λ *separate the points of* G.

□

COROLLARY 3. *Let* G *be Abelian. Then the system of its characters* $\{\chi_\lambda\}$ *is an orthogonal basis in* $L_2(G)$ *and is complete in* $L_1(G)$ *and* $C(G)$. *In these spaces the Fourier decomposition is a resolution of identity that reduces the regular representation. The same holds true in every space* $L_p(G)$ *with* $1 < p < \infty$.

4°. The classical theory of representations of finite groups is covered by the Peter-Weyl theory. We summarize its main results in the following theorem.

THEOREM. *Any finite group* G *has only a finite set of pairwise nonequivalent irreducible representations* $V_1,\ldots,V_r$. *They are finite dimensional and can be chosen unitary. Let* n_λ, $\tau_{\lambda,ik}$ $(1 \leqslant i,k \leqslant n_\lambda)$, *and* χ_λ *denote respectively the degree, matrix elements, and characters of the representation* V_λ, $\lambda = 1,\ldots,r$. *Then* :

1) *the following orthogonality relations hold for the matrix elements* :

$$\frac{1}{|G|} \sum_g \tau_{\lambda,ik}(g)\overline{\tau_{\mu,j\ell}(g)} = \frac{1}{n_\lambda} \delta_{\lambda\mu}\delta_{ij}\delta_{k\ell} ,$$

and for the characters :

$$\frac{1}{|G|} \sum_g \chi_\lambda(g)\overline{\chi_\mu(g)} = \delta_{\lambda\mu} ;$$

2) *the system of matrix elements is a basis in the space of all functions on* G;

3) *the system of characters is a basis in the space of all functions on the set* $\tilde{G}$ *of conjugacy classes of* G.

□

From among the numerous consequences of this theorem we mention two combinatorial results.

COROLLARY 1. *The number of pairwise nonequivalent irreducible representations of a finite group* G *is equal to the number of conjugacy classes of* G.

In particular, if G is Abelian, then $|G^*| = |G|$.

COROLLARY 2 (Burnside's formula).

$$n_1^2 + \ldots + n_r^2 = |G| .$$

□

Every finite dimensional representation T of the finite group G decomposes (uniquely, to within equivalence) into an orthogonal direct sum $T = d_1V_1 \oplus \ldots \oplus d_rV_r$, with multiplicities $d_1,\ldots,d_r$. Accordingly, $\chi_T = d_1\chi_1 + \ldots + d_r\chi_r$. The orthogonality relations permit us to calculate the multiplicities d_λ as

$$d_\lambda = (\chi_T,\chi_\lambda) , \quad 1 \leqslant \lambda \leqslant r .$$

2. BANACH REPRESENTATIONS OF COMPACT GROUPS AND SEMIGROUPS

1°. The following general result is directly related to the Peter-Weyl theory.

THEOREM. *Given any representation* T *of a compact group* G *in a Banach space* B, *there exists a resolution of identity* $\{P_\lambda\}$ *which reduces* T *and has the following properties* :

1) If $x \in \text{Im}\, P_\lambda$, $x \neq 0$, *then the linear span of the orbit* O(x) *is finite dimensional and decomposes into a direct sum of at most* n_λ *invariant subspaces such that the restriction of* T *to each of these subspaces is equivalent to an irreducible representation* V_λ *of* G.

2) Every invariant subspace on which T *is equivalent to* V_λ *is contained in* $\text{Im}\, P_\lambda$.

The Peter-Weyl L_1- and C-Theorems are covered by this formulation. However, our proof will use the L_1-theorem.

PROOF. Preserving the notations of the preceding section, we consider the operators $P_\lambda = \tilde{\pi}_\lambda$, the Fourier transforms of the functions π_λ under representation T. Explicitly,

$$P_\lambda x = n_\lambda \int \chi_\lambda(g) T(g^{-1}) x \, dg \quad , \quad x \in B,$$

(recall that $\pi_\lambda = n_\lambda \chi_\lambda$). Since the Fourier transformation is a morphism of the group algebra $L(G;T) \approx L_1(G)$ into $L(B)$,

$$P_\lambda P_\mu = \delta_{\lambda\mu} P_\mu \ ,$$

i.e., $\{P_\lambda\}$ is an algebraically-orthogonal family of projections. Each P_λ commutes with the representation T ; in fact, P_λ is the Fourier transform of a central function, and the operators T(g) belong to the strong closure of the Fourier image of the group algebra.

For each vector $x \in B$ we let $F_x : L_1(G) \to B$ denote the continuous homomorphism defined by the rule $F_x\phi = \tilde{\phi}x$. It intertwines the regular representation R and representation T : $F_xR(g) = T(g)F_x$ for all $g \in G$. Also, $F_x\pi_\lambda = P_\lambda x$.

To prove the completeness of the family of projections $\{P_\lambda\}$, suppose that the functional $f \in B^*$ annihilates $\operatorname{Im} P_\lambda$ for all λ. Let $x \in B$ and $\phi \in L_1(G)$. Since $F_x(\pi_\lambda * \phi) = P_\lambda F_x\phi$, the functional F_x^*f given by $(F_x^*f)(\psi) = f(F_x\psi)$ annihilates $\operatorname{Im} \Pi_\lambda$ in $L_1(G)$ for all λ. Consequently, $F_x^*f = 0$, i.e., $f|\operatorname{Im} F_x = 0$. But $x \in \overline{\operatorname{Im} F_x}$. Hence, $f(x) = 0$, and since x is arbitrary, we conclude that $f = 0$.

To establish the totality of the family $\{P_\lambda\}$, let $x \in \cap_\lambda \operatorname{Ker} P_\lambda$. Since $F_x(\pi_\lambda * \phi) \equiv 0$, we have $F_x^*f = 0$ for all $f \in B^*$, i.e., $F_x = 0$. Since $x \in \overline{\operatorname{Im} F_x}$, we conclude that $x = 0$.

Thus, $\{P_\lambda\}$ is a resolution of identity that reduces the given representation T.

Now let $x \in \operatorname{Im} P_\lambda$, $x \neq 0$. Then $x = P_\lambda x = F_x\pi_\lambda$. Since the linear span of the orbit of the function π_λ under the regular representation is M_λ (see page 130), with $\dim M_\lambda = n_\lambda^2$, it follows that $\dim \operatorname{Lin} O(x) \leqslant n_\lambda^2$. Decomposing M_λ into a direct sum of n_λ invariant subspaces, in each of which the regular representation R is equivalent to V_λ, we obtain a decomposition of $\operatorname{Lin} O(x)$ into a sum of n_λ invariant subspaces, in each of which the representation T is intertwined with V_λ. Those of these subspaces on which T is not equivalent to V_λ reduce to zero by the Second Schur Lemma. The remaining subspaces are not necessarily independent, but one can isolate an independent subfamily without affecting their sum. This yields the needed decomposition for $\operatorname{Lin} O(x)$.

Finally, let L be an invariant subspace such that $T|L \sim V_\lambda$. Then every generalized matrix element $\phi(g) = f(T(g)x)$ $(x \in B,$ $f \in B^*)$ belongs to M_λ. Consequently, $f(P_\lambda x) = (\pi_\lambda * \phi)(e) =$ $= \phi(e) = f(x)$. Since f is arbitrary, we get $x = P_\lambda x$ for all $x \in L$, i.e., $L \subset \operatorname{Im} P_\lambda$.

□

COROLLARY 1. *The family of all finite dimensional invariant subspaces of any Banach representation of a compact group is complete.*

□

COROLLARY 2. *Every irreducible Banach representation of a compact group is finite dimensional.*

□

In particular, *every such representation is equivalent to a unitary representation.*

COROLLARY 3. *Every irreducible Banach representation of a compact Abelian group is one-dimensional.*

□

Thus, such representations can be identified with the unitary characters of the group.

The family of projections $\{P_\lambda\}$ is uniquely specified by the properties 1) and 2) in the statement of the theorem. In fact, the image $\operatorname{Im} P_\lambda$ is equal to the set of all vectors satisfying 1). This uniquely specifies all $\operatorname{Im} P_\lambda$, and $\operatorname{Ker} P_\lambda = \sum_{\mu \neq \lambda} \operatorname{Im} P_\mu$.

Those subspaces $\operatorname{Im} P_\lambda$ which are different from zero are called the *primary* or *isotypical components* of representation T. In the Abelian case the primary components coincide with the corresponding weight subspaces. Corollary 1 then shows that *every Banach representation of a compact Abelian group possesses a complete system of weight vectors.*

Exercise 1. *If the compact group G is Abelian and representation T is isometric, then all P_λ are orthogonal projections. In other words, the resolution of identity $\{P_\lambda\}$ is in this case orthogonal.* [Recall that the representation can be always made isometric by replacing the norm in B with an equivalent norm.]

Exercise 2. *Let τ be a continuous homomorphism of the compact Abelian group G into the group of invertible elements of a commutative Banach algebra. Then $\tau(g) = \sum_{k=1}^{n} \chi_k(g) p_k$ (for all $g \in G$), where $\{p_k\}_1^n$ is a set of mutually annihilating* *idempotents satisfying* $\sum_{k=1}^{n} p_k = e$ *(the identity element of the algebra) and* χ_k *are unitary characters of* G. *In particular, if the maximal ideal space of the algebra is connected, then* $\tau(g) = \chi(g)e$, *where* χ *is a unitary character of* G (E. A. Gorin, 1970).

2°. We now turn to representations of compact semigroups. The Kernel Theorem permits us to reduce this case to that of groups.

Let S be a compact semigroup of type 1×1 (in particular, S is Abelian). Let K be the Sushkevich kernel of S and e the identity element of K. The canonical retraction $\hat{e} : S \to K$ ($\hat{e}s = se$) enables us to extend every representation T of the compact group K to a representation of S by the rule : $T(s) = T(\hat{e}s) = T(se)$. The representations of the semigroup S arising in this manner will be termed *nondegenerate*. Representation T is nondegenerate if and only if $T(e) = E$. The necessity of this condition is obvious. Its sufficiency follows from the chain of equalities $T(s) = T(s)E = T(s)T(e) = T(se)$. In the general case the operator $P = T(e)$ is a projection, since e is an idempotent. We call P the *boundary operator* of representation T. If P is injective, then $P = E$ and T is nondegenerate. Conversely, if T is nondegenerate, then all operators $T(s)$ are injective (and even invertible). Here we should emphasize that for a nondegenerate representation T the operator semigroup $\{T(s)\}_{s \in S}$ coincides with the group $\{T(s)\}_{s \in K}$. The preceding theorem applies to every nondegenerate representation T, i.e., every such T decomposes into primary components corresponding to irreducible representations of the kernel K. This decomposition is orthogonal whenever T is contractive (in particular, isometric) and S is Abelian or the representation space B is Hilbert.

THEOREM. *Let* T *be a representation of the compact semigroup* S *of type* 1×1 *in the Banach space* B. *Let* T_K *denote the restriction of* T *to the Sushkevich kernel* K *of* S. *Then* $B = B_1 \dot{+} B_0$, *where* B_1 *and* B_0 *are* T-*invariant subspaces such that*

1) the representation $T|B_1$ *is nondegenerate and* B_1 *is the largest* T*-invariant subspace with this property ;*

2) $T_K | B_0 = 0$ *and* $\operatorname{Ker} T(s) \subset B_0$ *for all* $s \in S$ *(so that* $B_0 = \cup_s \operatorname{Ker} T(s)$ *and* $\operatorname{Ker} T(s) = B_0$ *for all* $s \in K$*).*

[Notice that T_K is, generally speaking, only a semigroup representation of the group K, i.e., $T_K(e) \neq E$.]

PROOF. Consider the projection $P = T(e)$. Since e commutes with every $s \in S$, P commutes with representation T. We put $B_1 = \operatorname{Im} P$ and $B_0 = \operatorname{Ker} P$. Then the representation $T|B_1$ is nondegenerate because $T(e)|B_1 = P|B_1 = \mathrm{id}$. Conversely, if L is an invariant subspace with the property that $T|L$ is nondegenerate, then $T(e)|L = \mathrm{id}$, i.e., $P|L = \mathrm{id}$, and so $L \subset B_1$. Now let $s \quad K$. Then $s = se$, which gives $T(s) = T(s)P$. Consequently, $T(s)|B_0 = T(s)P|\operatorname{Ker} P = 0$. If now $T(s)x = 0$ for some $s \in S$, then $T(s)Px = PT(s)x = 0$. But $Px \in B_1$ and $T(s)|B_1$ is invertible. Therefore, $Px = 0$, i.e., $x \in B_0$, showing that $\operatorname{Ker} T(s) \subset B_0$ for all $s \in S$.

□

The subspaces B_0 and B_1 are obviously uniquely specified by properties 1) and 2). Furthermore, B_1 is equal to the closure of the sum of all finite dimensional invariant subspaces on which T is equivalent to an irreducible unitary representation (in the Abelian case B_1 coincides with the closure of the sum of the weight subspaces corresponding to unitary weights). We call B_1 and B_0 the *boundary* and the *interior* subspaces of representation T.

The theorem just proved extends to the case of a real Banach space B. On complexifying a real representation T, the complexified boundary and interior subspaces of T become the corresponding subspaces of the complexified T. This observation will be used in Sec. 4.

<u>Remark.</u> *Suppose that in the preceding theorem* T *is a contractive representation. Then the boundary projection* P *is orthogonal (and so the decomposition* $B = B_1 \dotplus B_0$ *is semiortho-*

gonal), and the representation $T|B_1$ *is isometric, with all the ensuing consequences.* In fact, in this case $\|T(e)\| \leq 1$, i.e., $\|P\| \leq 1$, and $T|B_1$, being a contractive representation of the group K, is isometric.

Exercise. *Let* V *be an irreducible representation of the compact semigroup* S *of type* 1×1. *Suppose there is a vector* x *such that* $V(s)x \neq 0$ *for all* $s \in S$. *Then* V *is finite dimensional and equivalent to a unitary representation.*

3. ALMOST PERIODIC REPRESENTATIONS AND FUNCTIONS

1°. The theory of representations of compact groups gives a remarkably transparent method of constructing the theory of almost periodic functions (a.p.f.'s), created by H. Bohr in 1925. [Bohr had predecesors : the dissertations of P. Bohl (1893) and E. Esclangon (1904) studied the less wide, yet important class of *quasi-periodic* functions. Bohr gave a systematic treatment of his theory in the monograph [3].]

The simplest example of an a.p.f. (leaving aside the periodic functions) is $\sin t + \sin \omega t$ $(t \in \mathbb{R})$, where ω is an irrational number. Such oscillations are known in physics as *beats*. The general contruction of a.p.f.'s amounts to taking the uniform closure, over the full real axis, of the linear span of all exponential $e^{i\lambda t}$ with $\lambda \in \mathbb{R}$. On the other hand, a.p.f.'s can be defined intrinsically. Specifically, a continuous function $\phi(t)$, $t \in \mathbb{R}$, is said to be *almost periodic (a.p.)* if for every $\varepsilon > 0$ the set of its ε-almost periods is relatively dense (by definition, τ is an ε-*almost period* of the function ϕ if $\sup_t |\phi(t+\tau) - \phi(t)| < \varepsilon$; also, a set $M \subset \mathbb{R}$ is said to be *relatively dense* if there is an $\ell > 0$ such that every interval $(a, a+\ell)$ intersects M). The basic approximation theorem of Bohr establishes the equivalence of these two definitions of the class of a.p.f.'s. A third equivalent definition was proposed by S. Bochner in 1927 : a function $\phi \in CB(\mathbb{R})$ is *almost periodic* if the family of its translates

$\{\phi(t+\tau)\}_{\tau \in \mathbb{R}}$ is precompact in $CB(\mathbb{R})$. This turned out to be a rather felicitous definition, since it led to the discovery of an intimate connection between the theory of a.p.f.'s and the Peter-Weyl theory (H. Weyl, 1927; L. S. Pontryagin, 1933) and opened the way to a wide range of generalizations (J. von Neumann, 1934). The modern outcome of Bochner's approach is the theory of a.p. representations.

Let T be a bounded representation of a topological semigroup S in a Banach space B. We call $x \in B$ an *almost periodic vector* (*a.p.v.*) if the orbit $O(x)$ is precompact. The set of all a.p.v.'s of representation T will be denoted by $(AP)_T$.

Exercise. $(AP)_T$ *is an invariant subspace.*

A bounded representation T is said to be *almost periodic* (*a.p.*) if $(AP)_T = B$, i.e., every $x \in B$ is an a.p.v. This is a variant of the definition of an a.p. operator semigroup proposed by K. de Leeuw and I. Glicksberg in 1961 (their work, of which we will make essential use in the sequel, was preceded by fundamental investigations of W. Maak (1954) and K. Jacobs (1956, 1957)). An *a.p. semigroup of operators* in B is defined as a subsemigroup of End B whose trivial representation is a.p. .

Remark. We could have defined an a.p. representation as one for which all orbits $O(x)$ are precompact, without requiring boundedness beforehand. Under this definition boundedness is guaranteed automatically.

Exercise 1. *In a finite dimensional space every bounded representation is a.p. .*

Exercise 2. *Let* T *be a bounded representation. Then* $T|(AP)_T$ *is a.p. (this is the largest a.p. subrepresentation of* T*).*

Exercise 3. *Let* x *be an a.p.v. of the bounded representation* T*. Then the representation* $T|\overline{\text{Lin}\, O(x)}$ *is a.p. .*

Exercise 4. *Suppose* T *is a bounded irreducible representation and there exists an a.p.v.* $x \neq 0$. *Then* T *is a.p.* .

Exercise 5. *Every subrepresentation of an a.p. representation is a.p.* .

We next build bridges between the theory of a.p. representations and the theory of representations of compact groups. In one direction this is a rather easy task.

LEMMA. *Every representation of a compact semigroup is a.p.* .

In fact, each orbit of such a representation is compact as the image of a compact space under a continuous map.

□

The key to the passage in the opposite direction is the following result.

LEMMA (K. de Leeuw - I. Glicksberg, 1961). *Representation* T *is a.p. if and only if the closure of* $\operatorname{Im} T$ *is compact in the strong operator topology.*

PROOF. NECESSITY. Consider the compact set $Q = \prod_x \overline{O(x)}$ and the natural map $\Delta : \overline{\operatorname{Im} T} \to Q$ defined by the rule : $(\Delta A)(x) = Ax$. Then Δ is a homeomorphism of $\overline{\operatorname{Im} T}$ onto its image ; in fact, Δ is obviously injective, and the strong topology on $\overline{\operatorname{Im} T}$ is equivalent, by definition, to the product topology. It now suffices to show that $\Delta(\overline{\operatorname{Im} T})$ is closed in Q. But any limit point a of $\Delta(\overline{\operatorname{Im} T})$ specifies an operator $A \in L(B)$ thanks to the strong closedness of $L(B)$ in the space of all maps $B \to B$. Moreover, $A \in \overline{\overline{\operatorname{Im} T}} = \overline{\operatorname{Im} T}$, and clearly $a = \Delta A$.

SUFFICIENCY. Fix x and consider the continuous map $A \to Ax$ from $L(B)$, endowed with the strong operator topology, to B. It takes the set $\overline{\operatorname{Im} T}$, which by hypothesis is compact, into a compact set containing $O(x)$.

□

Let T be an a.p. representation of the semigroup S. We call $\beta(T) \equiv \overline{\mathrm{Im}\, T}$ the *Bohr compactum* of T (in the honor of H. Bohr ; he did not introduce this notion, which gradually emerged in later works). [Transl. note : here we follow the terminology of the author and use *compactum* rather than *compactification*, the term customarily used in the literature.]

LEMMA. *The Bohr compactum* $\beta(T)$ *of the a.p. representation* T *is a compact semigroup relative to multiplication of operators, and a compact group if* S *is a group.*

PROOF. Since the set $\beta(T) \subset L(B)$ is bounded, multiplication is continuous on $\beta(T)$ in the strong operator topology. Moreover, $\beta(T)$ is closed under multiplication, because such is its dense subset Im T. Thus, $\beta(T)$ is a compact semigroup. If S is a group, then such is Im T. Consequently, the group Γ of invertible elements of the semigroup $\beta(T)$ is dense in $\beta(T)$. Since, by a known general theorem, Γ is a compact group, it follows that $\Gamma = \beta(T)$, and hence that $\beta(T)$ is a compact group.

□

The Sushkevich kernel K of the compact semigroup $\beta(T)$ and its type are called the *Sushkevich kernel* and respectively the *type of representation* T. The definitions of the notions of Bohr compactum, Sushkevich kernel, and type extend immediately to an a.p.v. x of an arbitrary bounded representation T via the subrepresentation $T|\overline{\mathrm{Lin}\, O(x)}$.

Exercise. *Let* T *be an a.p. representation and let* T_1 *be a subrepresentation of* T. *The Bohr compactum* $\beta(T_1)$ *is a quotient semigroup of the Bohr compactum* $\beta(T)$.

It is clear that the Bohr compactum of any a.p. representation of an Abelian semigroup S is Abelian.

2°. Let us apply the constructions described above to the regular representation of a topological semigroup S. A function $\phi \in CB(S)$ is called *right* (*left*) *a.p.* if the family of its right translates $\phi_t(s) = \phi(st)$, $t \in S$ (respectively, left translates ${}_t\phi(s) = \phi(ts)$, $t \in S$) is precompact in $CB(S)$.

LEMMA (W. Maak, 1938). *Right and left almost periodicity are equivalent.*

PROOF. Suppose the function ϕ is right a.p. and (with no loss of generality) real-valued. Given any $\varepsilon > 0$, we choose in the family of right translates of ϕ an ε-mesh $\phi_{t_1},\dots,\phi_{t_p}$. Next, we divide the segment $-\|\phi\| \leqslant \zeta \leqslant \|\phi\|$ into subsegments $I_1,\dots,I_m$ of lenghts less than ε, and we introduce the Lebesgue sets $\Lambda_{k,j} = \{s \mid \phi_{t_k}(s) \in I_j\}$. Obviously, $\cup_{j=1}^{m}\Lambda_{k,j} =$. Consider all possible intersections $\Lambda^{j_1,\dots,j_p} = \Lambda_{1,j_1} \cap \dots \cap \Lambda_{p,j_p}$. We have $\cup_{j_1,\dots,j_p} \Lambda^{j_1,\dots,j_p} = S$. The oscillation of each of the functions $\phi_{t_1},\dots,\phi_{t_p}$ on any of the sets $\Lambda^{j_1,\dots,j_p}$ is smaller than ε, because the oscillation of ϕ_{t_k} on $\Lambda_{k,j}$ is smaller than ε by the definition of $\Lambda_{k,j}$. Therefore,

$$|\phi_{t_k}(s) - \phi_{t_k}(s')| < \varepsilon\ , \quad s,s' \in \Lambda^{j_1,\dots,j_p}\ .$$

But then $|\phi_t(s) - \phi_t(s')| < 3\varepsilon$ for all $s,s' \in \Lambda^{j_1,\dots,j_p}$ and all $t \in S$, i.e., $|\phi(st) - \phi(s't)| < 3\varepsilon$ for all s,s' $\Lambda^{j_1,\dots,j_p}$ and all $t \in S$, or, equivalently, $\|{}_s\phi - {}_{s'}\phi\| < 3\varepsilon$ for all $s,s' \in \Lambda^{j_1,\dots,j_p}$. Now pick an arbitrary point in each $\Lambda^{j_1,\dots,j_p}$ (if this set is not empty). By the previuos inequalities, the left translates of ϕ corresponding to these points form a 3ε-mesh in the family of all left translates of ϕ. Hence, ϕ is left a.p.

□

We can now speak of a.p. functions on S omitting the prefix "right" or "left". *If the semigroup* S *is compact, then every continuous function on* S *is a.p.* .

<u>Exercise 1.</u> *The set* $AP(S)$ *of almost periodic functions on*

the semigroup S *is a closed translation-invariant subalgebra of the Banach algebra* CB(S).

Exercise 2. $\phi \in AP(S)$ *is invertible in this algebra if and only if* $\inf_s |\phi(s)| > 0$.

LEMMA. *Every a.p. function is biuniformly continuous.*

PROOF. Let $\phi_{t_1}, \ldots, \phi_{t_p}$ be an ε-mesh in the family of right translates $\{\phi_t\}_{t \in S}$ of the a.p.f. ϕ. It follows from the continuity of ϕ at the points $st_1, \ldots, st_p$ that every point $s \in S$ has a neighborhood N such that if $r \in N$, then $|\phi(rt_k) - \phi(st_k)| < \varepsilon$ for $1 \leqslant k \leqslant p$. At the same time, given any $t \in S$ there is a t_k such that $|\phi(rt) - \phi(rt_k)| < \varepsilon$ for all $r \in S$. Consequently, $|\phi(rt) - \phi(st)| < 3\varepsilon$ for all $r \in N$ and all $t \in S$, i.e., $\|{}_r\phi - {}_s\phi\| < 3$. Thus, ϕ is left uniformly continuous. The right uniform continuity of ϕ is established in the same manner using the left translates ${}_t\phi$ of ϕ.

□

In view of this lemma, a right regular representation R of the semigroup S is defined in the Banach space AP(S). By the definition of a.p.f.'s, R is an a.p. representation.

From now on we shall assume, unless otherwise stipulated, that S possesses an identity element e.

The Bohr compactum $\beta(R)$ is called the *Bohr compactum of the semigroup* S, denoted $\beta(S)$. Its Sushkevich kernel and its type are called the *Sushkevich kernel* and respectively the *type* of S.

Remark. To define the notions of Bohr compactum, Sushkevich kernel, and type for semigroups without an identity element one first adjoins such an element as an isolated point.

Exercise. *If* S *is a compact semigroup, then the Bohr compactum of* S *is topologically isomorphic to* S. *Therefore, for compact semigroups the notions of kernel and type retain their original meaning.*

The map R may be regarded as a continuous homomorphism of the semigroup S into its Bohr compactum (S), with the property that $R(e) = E$. In this quality R is called the *canonical homomorphism* $S \to \beta(S)$. It induces a homomorphism $R^* : C(\beta(S)) \to CB(S)$ of the algebras of continuous functions, defined by the rule : $(R^*\psi)(s) = \psi(R(s))$.

THEOREM. R^* *is a Banach algebra isometry of* $C(\beta(S))$ *onto* $AP(S)$.

Thus, the a.p.f.'s on S are canonically identified with continuous functions on the Bohr compactum $\beta(S)$. This is a key moment in the theory of a.p.f.'s, which now becomes the theory of continuous functions on a compact semigroup and is thereby reduced, from the viewpoint of harmonic analysis, to the Peter-Weyl theory (for groups or semigroups of type 1×1).

PROOF. We have to check that $\mathrm{Im}\, R^* = AP(S)$. That $\mathrm{Im}\, R^* \subset AP(S)$ is plain in view of the almost periodicity of the continuous functions on S. Now let $\phi \in AP(S)$. Notice that $\phi(s) = (R(s)\phi)(e)$ and that the evaluation map at e is a linear functional on $AP(S)$. Let $t \in \beta(S)$, i.e., t is an endomorphism of the space $AP(S)$ which belongs to the strong closure of the family $\mathrm{Im}\, R$. The rule: $\psi(t) = (t\phi)(e)$ defines a continuous function on $\beta(S)$, and we have $(R^*\psi)(s) = \psi(R(s)) = (R(s)\phi)(e) = \phi(e)$. That R^* is an isometry follows from the fact that $\mathrm{Im}\, R$ is dense in $\beta(S)$.

□

COROLLARY 1. *Every a.p.f. on* S *is a function on* $\mathrm{Im}\, R$ *(i.e., depends only on* $R(s)$*) and extends by continuity from* $\mathrm{Im}\, R$ *to the Bohr compactum* $\beta(S)$.

□

We call this extension the *Bohr extension* of functions. It can be identified with the Gelfand representation of algebra $AP(S)$. In fact, we have

COROLLARY 2. *The Bohr compactum* $\beta(S)$ *of* S *is homeomorphic to the maximal ideal space of the Banach algebra* $AP(S)$ *(and* R^{**}

serves as the canonical homeomorphism).

□

In point of fact, we could define the Bohr compactum $\beta(S)$ as the maximal ideal space $M(AP(S))$, but then some extra work would be required in order to introduce the multiplication operation on $\beta(S)$. The path we followed seems easier.

<u>Exercise.</u> *The Bohr extension of the function* ϕ *commutes with translations.*

Now let us show that the correspondence $S \to \beta(S)$ can be extended to a covariant functor from the category of topological semigroups (with identity) into the category of compact semigroups.

THEOREM. *Let* $f : S_1 \to S_2$ *be a morphism of topological semigroups. Then there exists a unique morphism* $\beta(f)$ *of the corresponding Bohr compacta such that the diagram*

$$\begin{array}{ccc} S_1 & \xrightarrow{f} & S_2 \\ {\scriptstyle R_1}\downarrow & & \downarrow{\scriptstyle R_2} \\ \beta(S_1) & \xrightarrow{\beta(f)} & \beta(S_2) \end{array}$$

where R_1, R_2 *are the canonical homomorphisms, commutes.*

PROOF. The morphism f induces the homomorphism $f^* : AP(S_2) \to AP(S_1)$, and hence a continuous map $\beta(f) : \beta(S_1) \to \beta(S_2)$ of the corresponding maximal ideal spaces. By definition, we have the commutative diagram

$$\begin{array}{ccc} AP(S_1) & \xleftarrow{\quad f^* \quad} & AP(S_2) \\ {\scriptstyle R_1^*}\uparrow & & \uparrow{\scriptstyle R_2^*} \\ C(\beta(S_1)) & \xleftarrow{\quad \beta(f)^* \quad} & C(\beta(S_2)) \end{array}$$

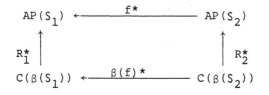

and so $[\beta(f)R_1]^* = (R_2 f)^*$. But then $\beta(f)R_1 = R_2 f$, too, since the continuous functions on the compactum $\beta(S_2)$ separate its points. It remains to verify that $\beta(f) : \beta(S_1) \to \beta(S_2)$ is a semigroup homomorphism. Let $t, t' \in \operatorname{Im} R_1$. Then $\beta(f)(tt') = \beta(f)t \cdot \beta(f)t'$, since R_1, R_2, and f are homomorphisms. This equality extends by continuity to arbitrary $t, t' \in \beta(S_1)$ thanks to the fact that $\operatorname{Im} R_1$ is dense. The latter is also the obvious reason for the uniqueness of the morphism $\beta(f)$.

□

COROLLARY 1. *The correspondence* $S \to \beta(S)$, $f \to \beta(f)$, *is a covariant functor from the category of topological semigroups with identity into the category of topological semigroups.*

□

We call it the *Bohr functor.*

COROLLARY 2. *Let* H *be a compact semigroup (with identity) and let* $f : S \to H$ *be a morphism. Then there exists a unique morphism* $\bar{f} : \beta(S) \to H$ *such that the diagram*

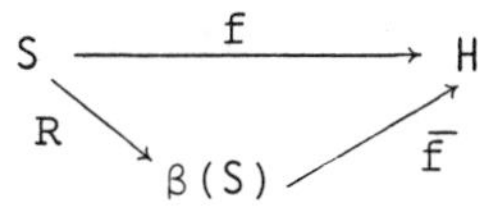

where R *is the canonical homomorphism, commutes.*

PROOF. Put $\bar{f} = h\beta(f)$, where h denotes the inverse of the canonical homomorphism $H \to \beta(H)$, which is a topological isomorphism thanks to the compactness of H. The uniqueness of $\bar{f}$ is guaranteed by the fact that $\operatorname{Im} R$ is dense in $\beta(S)$.

□

We call the morphism $\bar{f}$ the *Bohr extension of the morphism* f.

Each a.p. representation T of the semigroup S may be regarded as a morphism of S into the Bohr compactum $\beta(T)$ of T. Applying the preceding corollary, we obtain

COROLLARY 3. *Every a.p. representation* T *of the semigroup* S *can be written as* $T = \bar{T}R$, *where* $\bar{T}$ *is a uniquely determined representation of the Bohr compactum* $\beta(S)$ *and* $R : S \to \beta(S)$ *is the canonical homomorphism.*

□

We call $\overline{T}$ the *Bohr extension of representation* T.

Conversely, given a representation $\tilde{T}$ of the Bohr compactum $\beta(S)$, $T = \tilde{T}R$ is an a.p. representation of the semigroup S, and $\overline{T} = \tilde{T}$.

These results completely reduce the theory of a.p. representations to the theory of representations of compact semigroups.

We remark also that the canonical homomorphism $R : S \to \beta(S)$ is not always injective. *A necessary and sufficient condition for the injectivity of* R *is that the a.p.f.'s on* S *should separate points.* In fact, if R is injective, then the condition follows from the fact that the continuous functions on $\beta(S)$ separate points. Conversely, suppose that the a.p.f.'s on S separate points. Then if $s, t \in S$, $s \neq t$, there exists an a.p.f. ϕ such that $\phi(s) \neq \phi(t)$, i.e., $(R(s)\phi)(e) \neq (R(t)\phi)(e)$, and so $R(s) \neq R(t)$.

On the other hand, *for* R *to be injective it is necessary and sufficient that the semigroup* S *admit a continuous monomorphism into a compact semigroup.* The necessity of this condition is plain ; its sufficiency follows from Corollary 2.

3°. Following the path outlined above, we apply the theory of representation of compact semigroups to the theory of a.p. representations.

BASIC THEOREM. *Let* S *be a topological semigroup (with identity) of type* 1×1, $R : S \to \beta(S)$ *the canonical homomorphism of* S *into its Bohr compactum* $\beta(S)$, *and* $r : \beta(S) \to K$ *the canonical retraction onto the Sushkevich kernel* K *of* $\beta(S)$. *Let* T *be an a.p. representation of* S *in a Banach space* B. *Then*

$$B = B_1 \dotplus B_0 ,$$

where B_1 *and* B_0 *are* T*-invariant subspaces with the following properties :*

1) $T|B_1 = \tilde{T}rR$, *where* $\tilde{T}$ *is a representation of the compact group* K. *Accordingly,* $T|B_1$ *decomposes into primary components corresponding to irreducible unitary representations of* K *(these components are also irreducible unitary representations for* S*);*

2) the closure of the orbit $O(x)$ *contains the zero vector for every* $x \in B_0$;

3) B_1 *and* B_0 *are the largest* T-*invariant subspaces with the indicated properties.*

If T *is contractive, then the decomposition* $B = B_1 \dotplus B_0$ *is semiorthogonal and the representation* $T|B_1$ *is isometric. Therefore, these properties can be guaranteed by replacing the norm on* B *with an equivalent norm.*

[This result is a variant of a theorem proved by K. de Leeuw and I. Glicksberg (1961) for "weakly almost periodic" operator semigroups.]

The terminology used earlier for the subspaces B_1, B_0, and the associated projection P is preserved. If S is Abelian, the discrete spectrum of the representation $T|B_1$ is called the *boundary spectrum of* T. [In connection with this the Basic Theorem may be referred to as the *the Boundary Spectrum Splitting--Off Theorem.*]

Although the theorem becomes almost evident after the preparations made above, we wish to comment on the main steps in its proof.

The first step is to pass from the given representation T to its Bohr extension $\overline{T}$: $T = \overline{T}R$. Being a representation of a compact semigroup of type *1 × 1*, $\overline{T}$ splits : $B = B_1 \dotplus B_0$, with $\overline{T}|B_1$ nondegenerate and $B_0 = \cup_{t\in\beta(S)} \operatorname{Ker} \overline{T}(t)$. We now have : 1) $\overline{T}|B_1 = \tilde{T}r$, where $\tilde{T}$ is a representation of the Sushkevich kernel, with all ensuing properties ; 2) if $x \in B_0$, then $\overline{T}(t)x = 0$ for some $t \in \beta(S)$. Since $\operatorname{Im} R$ is dense in $\beta(S)$, it follows that $0 \in \overline{\{ \overline{T}(R(s))x \}} = \overline{O(x)}$; 3) if L is a T-invariant subspace with the same properties as B_1, then $\overline{T}|L$ is nondegenerate, and consequently $L \subset B_1$. Now let x be a vector such that $0 \in \overline{O(x)}$. Suppose $\overline{T}(t)x \neq 0$ for all $t \in \beta(S)$. Then $\inf_t \|\overline{T}(t)x\| > 0$ thanks to the compactness of $\beta(S)$, and then also $\inf_s \|\overline{T}(R(s))x\| > 0$, i.e., $\inf_s \|T(s)x\| > 0$, contrary to our assumption. The supplementary assertions regarding contractive representations require no special explanations.

□

The Abelian case deserves special consideration not only in view of its already familiar specific properties (guarantee for type *1 × 1*, coincidence of the primary and weight components, orthogonality of the decomposition into primary components), but also because in this case property 2) can be considerably sharpened.

Every Abelian semigroup S can be turned into a directed set upon taking the divisibility relation as a quasi-order : $s \geqslant t \Leftrightarrow \exists v : s = tv$. Obviously, $s_1 s_2 \geqslant s_1$ and $s_1 s_2 \geqslant s_2$ for all $s_1, s_2 \in S$. Let T be a contractive representation of S. Then $s \geqslant t$ implies $\|T(s)x\| \leqslant \|T(t)x\|$ for all x. Therefore, $\lim_s \|T(s)x\|$ exists. If $0 \in \overline{\{T(s)x\}}$, then obviously $\lim_s \|T(s)x\| = 0$, i.e., the orbit of x tends to zero. This fact is preserved on passing to an equivalent norm. Thus, if in the Basic Theorem S is Abelian, then $B_0 = \{x \mid \lim_s T(s)x = 0\}$.

Let us pause to discuss the case where S is a group. Then $\beta(S)$ is a compact group, $K = \beta(S)$, and $r = \text{id}$. Here $B_0 = 0$ and the Basic Theorem has the following

COROLLARY. *Let* T *be a Banach a.p. representation of the topological group* G. *Then* T *factors as* $T = \overline{T}R$, *where* $R : G \to \beta(G)$ *is the canonical homomorphism and* $\overline{T}$ *is a representation of the Bohr compactum* $\beta(G)$. *Accordingly,* T *decomposes into primary components corresponding to the irreducible unitary representations of* G.

If G *is Abelian, then the decomposition of* T *into primary components is orthogonal and the components themselves are weight subspaces.*

□

This formulation incorporates, in particular, the following important result : *the system of finite dimensional invariant subspaces of any a.p. representation* T *of the group* G *is complete* (on each of these subspaces the representation T can be considered irreducible, and hence unitary to within equivalence). This admits a converse : *if the bounded representation* T *of an arbitrary semigroup* S *possesses a complete system of finite dimensional invariant subspaces, then* T *is a.p.* . In fact, every $x \in B$ can be approximated by a sum of vectors $x_1, \ldots, x_n$ with finite dimensional orbits : $\|x - \sum_{k=1}^{n} x_k\| < \varepsilon$. Then $\|T(s)x - \sum_{k=1}^{n} T(s)x_k\| < c\varepsilon$ (with $c = \sup \|T(s)\|$), i.e., the distance from $O(x)$ to the set $\sum_{k=1}^{n} O(x_k)$ is less than ε. But a sum of precompact sets is precompact, and a set which can be arbitrarily well approximated by precompact sets is itself precompact.

In particular, *every a.p. representation of an Abelian group possesses a complete system of weight vectors.* Conversely, *if a bounded representation of a semigroup possesses a complete system of weight vectors, then it is a.p.* .

We shall come back later to this circle of problems. For the moment, let us formulate the main results of the theory of a.p.f.'s on groups which follow from the foregoing analysis.

THEOREM (J. von Neumann, 1934). *Let* G *be a topological group,* $\{V_\lambda\}$ *the system of all pairwise nonequivalent finite dimensional unitary representations of* G, *and* $\{\tau_{\lambda,ik}\}_{i,k=1}^{n_\lambda}$ *the corresponding system of matrix elements. Then all functions* $\tau_{\lambda,ik}$ *are a.p. and the full system* $\{\tau_{\lambda,ik}\}_{i,k,\lambda}$ *is complete in* $AP(G)$. *In particular, if* G *is Abelian, then* $AP(G)$ *is equal to the uniform closure of the system of unitary characters of* G.

To show this it suffices to apply the group variant of the Basic Theorem to the regular representation R of G. Following this path one can also develop the theory of Fourier series of a.p.f.'s.

LEMMA. *There exists a unique biinvariant mean on the space* $AP(G)$.

PROOF. For each a.p.f. ϕ we put

$$< \phi > = \int_{\beta(G)} \overline{\phi}(\gamma)d\gamma \ ,$$

where $\overline{\phi}$ is the Bohr extension of ϕ and $d\gamma$ is the normalized Haar measure on the Bohr compactum $\beta(G)$. Then $< \cdot >$ is obviously a biinvariant mean on $AP(G)$. To prove its uniqueness, suppose $[\cdot]$ is an invariant (even one-sided, say, for definiteness, right) mean on $AP(G)$. Then via the canonical isomorphism, $[\cdot]$ is also an invariant mean on $C(\beta(G))$. Therefore, we necessarily have $[\cdot] = < \cdot >$.

□

We can now correspond to each a.p.f. ϕ on G the *Fourier series*

$$\phi \sim \sum_\lambda \Pi_\lambda \phi \ ,$$

where

$$(\Pi_\lambda \phi)(h) = < \pi_\lambda(g)\phi(hg^{-1}) >_g$$

(as before, $\pi_\lambda = n_\lambda \chi_\lambda$, where χ_λ denotes the character of the representation V_λ). To within the Bohr extension, this is just the Fourier series of the function $\overline{\phi}$. By the Peter-Weyl L_2-Theorem, *the Fourier series of any a.p.f.* ϕ *converges to* ϕ *in the Hilbert metric defined by the norm* $\|\phi\|^2 = < |\phi|^2 >$. The latter is referred to as the B_2-*norm*. Moreover, *the generalized Parseval equality holds* :

$$\|\phi\|^2 = \sum_\lambda \|\Pi_\lambda \phi\|^2 \ .$$

A consequence of this is that there is an at most countable (and nonempty for $\phi \neq 0$) set of values λ for which $\Pi_\lambda \phi \neq 0$. This set is called the *Bohr spectrum of the a.p.f.* ϕ and is denoted by $\mathrm{spec}_B \phi$.

Exercise. *Any a.p.f. ϕ on the group G can be arbitrarily well approximated by linear combinations of matrix elements of those representations V_λ for which $\lambda \in \mathrm{spec}_B\phi$.*

In the Abelian case $\phi \sim \sum_\lambda c_\lambda \chi_\lambda$, where $c_\lambda = \langle \phi\chi_\lambda^* \rangle$ are called the *Fourier coefficients of* ϕ.

For the group $G = \mathbb{R}$ we obtain Bohr's theory. Here $\chi_\lambda(t) = e^{i\lambda t}$, $\lambda \in \mathbb{R}$, von Neumann's theorem becomes Bohr's approximation theorem, and the Fourier series of ϕ takes on the form

$$\phi \sim \sum_\lambda c_\lambda e^{i\lambda t} ,$$

where

$$c_\lambda = \langle \phi(t)e^{-i\lambda t} \rangle_t \quad .$$

Exercise. *The following formula of Bohr holds on the group* $\mathbb{R}$:

$$\langle\phi\rangle = \lim_{t_2-t_1\to\infty} \frac{1}{t_2-t_1}\int_{t_1}^{t_2} \phi(t)\,dt \;;$$

consequently,

$$c_\lambda = \lim_{t_2-t_1\to\infty} \frac{1}{t_2-t_1}\int_{t_1}^{t_2} \phi(t)e^{-i\lambda t}dt \; .$$

[Bohr's formula has been extended to locally compact Abelian groups by G. Ya. Lyubarskii (1948).]

At present the theory of almost periodic functions constitutes an important and far advanced domain of mathematical analysis. For further acquaintance with this theory we recommend the monograph [31] of B. M. Levitan. We wish nevertheless to discuss here some of its more important questions.

4°. The next result goes back to the very beginning of the theory of a.p.f.'s.

THEOREM (P. Bohl, 1906; H. Bohr, 1925). *Suppose the function*

$\phi(t)$ $(t \in \mathbb{R})$ *is bounded, differentiable, and its derivative* ϕ' *is a.p. . Then* ϕ *is also a.p. .*

□

This statement can be extended to arbitrary topological groups G as follows.

THEOREM (R. Doss, 1961). *Suppose the function* $\phi \in CB(G)$ *has the property that all its increments* $\Delta_h(g) = \phi(gh) - \phi(g)$ $(h \in G)$ *are a.p.f.'s. Then* ϕ *is also a.p. .*

In the Bohl-Bohr case, $\Delta_h(t) = \int_0^h \phi'(t+\tau)d\tau$, and the almost periodicity of all increments is an obvious consequence of the almost periodicity of ϕ'.

PROOF. Suppose that ϕ is (with no loss of generality) real-valued and is not a.p. . Then there is an $\varepsilon > 0$ such that the family of left translates ${}_t\phi$ of ϕ does not admit a finite ε-mesh, and so one can find a sequence $\{t_n\}_1^\infty$ such that $\inf_{n \neq m} \sup_g |\phi(t_n g) - \phi(t_m g)| > \varepsilon$. Let $\sup_{g,g'} |\phi(g') - \phi(g)| = M > 0$ (clearly, $M \geq 0$, and $M = 0$ implies $\phi = \text{const}$, which is excluded). Given $\eta \in (0,\varepsilon)$, we choose $g_1, g_0 \in G$ such that $\phi(g_1) - \phi(g_0) = M - \varepsilon + \eta$, i.e., $\Delta_h(g_0) = M - \varepsilon + \eta$, where $h = g_0^{-1}g_1$. Since Δ_h is an a.p.f., there exist $s_1,\ldots,s_k \in G$ such that for each $g \in G$ one can find an index i for which $|\Delta_h(gs_i) - \Delta_h(g_0)| < \frac{\eta}{2}$, and hence $\Delta_h(gs_i) > M - \varepsilon + \frac{\eta}{2}$. Thus, for every $g \in G$ there is an i, $1 \leq i \leq k$, such that $\phi(gs_ih) - \phi(gs_i) > M - \varepsilon + \frac{\eta}{2}$. Consider the collection of a.p.f.'s $\Delta_{s_ih}(g)$ and their left translates corresponding to the sequence $\{t_n\}$ constructed in the beginning of the proof. We may clearly assume that $\|{}_{t_n}\Delta_{s_ih} - {}_{t_m}\Delta_{s_ih}\| < \frac{\eta}{2}$, $1 \leq i \leq k$, i.e.,

$$|\phi(t_n gs_ih) - \phi(t_n g) - \phi(t_m gs_ih) + \phi(t_m g)| < \frac{\eta}{2}$$

for all $g \in G$ and $1 \leqslant i \leqslant k$. We fix a pair n,m, $n \neq m$, and find an element $g^0 \in G$ such that $|\phi(t_n g^0) - \phi(t_m g^0)| > \varepsilon$. Actually, we may assume that $\phi(t_n g^0) - \phi(t_m g^0) > \varepsilon$. Then $\phi(t_n g^0 s_i h) - \phi(t_m g^0 s_i h) > \varepsilon - \frac{\eta}{2}$ for $i = 1,\ldots,k$. Next, for the element $t_m g^0$ we choose an i such that $\Delta_h(t_m g^0 s_i) > M - \varepsilon + \frac{\eta}{2}$, i.e., $\phi(t_m g^0 s_i h) - \phi(t_m g^0 s_i) > M - \varepsilon + \frac{\eta}{2}$. But then $\phi(t_n g^0 s_i h) - \phi(t_m g^0 s_i) > M$, which contradicts the definition of M.

□

The Bohl-Bohr Theorem has been generalized in a different direction by S. Bochner (1933) and L. Amerio (1965), who considered the integral of functions $x(t)$ $(t \in \mathbb{R})$ with values in a Banach space. The definitive results in this direction belong to M. I. Kadets (1969), who found a condition on the space in which the functions assume values which is necessary and sufficient for the extendability of the Bohl-Bohr theorem [specifically, the condition is that there should be no subspaces isomorphic to the Banach space c of convergent sequences]. A syntesis of the theorems of Doss and Kadets has been carried out by R. Boles Basit (1971).

5°. Yet another interesting aspect of the theory of a.p.f.'s which in the final analysis also has a group nature is known as the *argument* (or *mean motion*) *theorem*, and goes back to the investigations of Lagrange in celestial mechanics.

THEOREM (H. Bohr, 1932; E.R. van Kampen, 1937). *Let* ϕ *be an a.p.f. on the connected group* G *such that* $\inf_g |\phi(g)| > 0$. *Then* $\phi(g) = \chi(g) \exp \psi(g)$, *where* χ *is an one-dimensional unitary character of* G *and* ψ *is an a.p.f.* .

□

E. A. Gorin and V. Ya. Lin (1967) discovered a connection between this fact and the topology of the Bohr compactum $\beta(G)$. Subsequently, E. A. Gorin (1970) generalized considerably the Bohr-van Kampen Theorem. Without touching upon this generalization itself, we present the elegant proof that it provide for the Bohr-van Kampen Theorem.

PROOF. The Bohr compactum $\beta(G)$ is connected, since the continuous image of the connected space G is dense in $\beta(G)$. The Bohr extension $\bar{\phi}$ of the given function ϕ is everywhere different from zero. Hence, it suffices to prove the theorem when G is a compact group and $\phi \in C(G)$. We may also assume that $\phi(e) = 1$. Set

$$\Theta(g,h) = \frac{\phi(gh)}{\phi(g)\phi(h)} \qquad (g,h \in G).$$

This defines a continuous function on the connected compact space $G \times G$; moreover, $\Theta(g,e) = 1$, $\Theta(e,h) = 1$, and $\Theta(g,h) \neq 0$ for all $(g,h) \in G \times G$. Under these circumstances it is readily established that $\Theta(g,h) = \exp \omega(g,h)$, where $\omega(g,h)$ is a continuous function satisfying $\omega(g,e) = 0$, $\omega(e,h) = 0$. Next, for any $g,h,k \in G$ we have

$$\omega(g,h) + \omega(gh,k) = \omega(g,hk) + \omega(h,k) .$$

Integrating this relation with respect to k (the Haar measure dk is normalized), we get

$$\omega(g,h) = \psi(gh) - \psi(g) - \psi(h) ,$$

where

$$\psi(g) = -\int \omega(g,k)\,dk .$$

This in turn yields

$$\frac{\exp \psi(gh)}{\exp \psi(g)\cdot\exp \psi(h)} = \frac{\phi(gh)}{\phi(g)\phi(h)} ,$$

which shows that the function $\chi = \phi\exp(-\psi)$ is a one-dimensional character.

□

For $G = \mathbb{R}$ we obtain $\phi(t) = \exp(i\lambda t + \psi(t))$, where ψ is an a.p. function, and so $\arg \phi(t) = \lambda t + \operatorname{im} \psi(t)$. A crude consequence of this formula is the existence of the *mean motion*

$$\lim_{t_2 - t_1 \to \infty} (t_2 - t_1)^{-1}[\arg \phi(t_2) - \arg \phi(t_1)] = \lambda.$$

Lagrange posed the problem of the existence of mean motion for an arbitrary ($\not\equiv 0$) linear combinations of exponentials. It was solved only in our century through efforts of prominent mathematicians (P. Bohl, 1909; H. Weyl, 1939; B. Jessen and H. Tornehave, 1945).

6°. We mention another "a.p. approach" to the completeness problem for the system of finite dimensional invariant subspaces. Let T be a representation of the topological semigroup S (with identity) in the Banach space B. We say that T is *scalarly-a.p.* if its generalized matrix elements $\tau_{f,x}(s) = f(T(s)x)$ $(x \in B,$ $f \in B^*)$ are a.p. functions on S.

Exercise 1. *Every a.p. representation is scalarly-a.p.* .

Exercise 2. *Every scalarly-a.p. representation is bounded.*

THEOREM. *Suppose* $S \equiv G$ *is a group and the space* B *is reflexive. Then every scalarly-a.p. representation of* G *in* B *is a.p.* .

PROOF. Consider the Bohr extensions $\overline{\tau_{f,x}}$ of the generalized matrix elements of T. Since the Bohr extension is isometric, $\|\overline{\tau_{f,x}}\| \leqslant c\,\|f\|\,\|x\|$, where $c = \sup_g \|T(g)\|$. Therefore, for fixed $t \in \beta(G)$ and $x \in B$, the map $f \to \overline{\tau_{f,x}}(t)$ is a linear functional of f. Since B is reflexive, $\overline{\tau_{f,x}} = f(\tilde{T}(t)x)$, where $\tilde{T}(t)$ is a map $B \to B$. It is readily verified that $\tilde{T}(t)$ is a linear operator and that $\|\tilde{T}(t)\| \leqslant c$. Setting $t = R(g)$ $(g \in G)$, where $R : G \to \beta(G)$ denotes the canonical homomorphism, we get

$$f(\tilde{T}(R(g))x) = \overline{\tau_{f,x}}(R(g)) = \tau_{f,x}(g) = f(T(g)x) \ .$$

Thus, $T = \tilde{T}R$, i.e., $\tilde{T}$ is a candidate for the role of the Bohr extension of representation T. We show that $\tilde{T} : \beta(G) \to L(B)$ is a representation, which implies that T is a.p.

It follows from the continuity of the functions $\overline{\tau_{f,x}}$ that the operator-valued function $\tilde{T}(t)$ is weakly continuous. Also, the operator-valued function $\tilde{T}(t_1t_2)$ is jointly weakly continuous in the variables t_1, t_2, while $\tilde{T}(t_1)\tilde{T}(t_2)$ is separately weakly continuous. For $t_1 = R(g_1)$ and $t_2 = R(g_2)$ we have $\tilde{T}(t_1t_2) = \tilde{T}(t_1)\tilde{T}(t_2)$, i.e., for fixed $t_1 = R(g_1)$ this equality holds for a dense subset of points $t_2 \in \beta(G)$, and hence for all t_2. Now fixing $t_2 \in \beta(G)$ we extend the equality by (weak) continuity to all $t_1 \in \beta(G)$. Moreover, $\tilde{T}(e) = E$. Therefore, $\tilde{T}$ is a weakly continuous homomorphism of the compact group $\beta(G)$ into Aut B. By a theorem already known to us, $\tilde{T}$ is a representation, as claimed.

□

COROLLARY. *Let* T *be a scalarly-a.p. representation of the topological group* G *in a reflexive Banach space. Then the system of invariant spaces of* T *is complete.*

□

For an Abelian group G this yields the completeness of the system of weight vectors. For $G = \mathbb{R}$ this result was obtained by the author in 1963.

COROLLARY (Yu. I. Lyubich, 1960). *Let* A *be a compact operator in the reflexive Banach space* B *such that* $\sup_{-\infty<t<\infty} \|e^{At}\| < \infty$. *Then the system of eigenvectors of* A *is complete.*

In fact, the representation $T(t) = e^{At}$ is scalarly-a.p., since the functions $f(e^{At}x)$, $f \in B^*$, are bounded, and their derivatives $f(Ae^{At}x)$ are a.p. thanks to the boundedness of T and the compactness of the operator A. The Bohl-Bohr Theorem used in this argument can be replaced by the Kadets Theorem, applied to the functions $t \to e^{At}x$ with values in B. The representation T is a.p. if B contains no subspaces isomorphic to c. Thus, if this last condition is satisfied, the system of eigenvectors of the operator A is complete.

□

Example (Yu. I. Lyubich, 1963). Let $\{\lambda_k\}_1^\infty \subset \mathbb{R}$, $\lambda_j \neq \lambda_k$ for $j \neq k$, and $\lim_{k\to\infty}\lambda_k = 0$. In c consider the operator A given by $A\{\xi_k\} = \{i\lambda_k \xi_k\}$. It is compact, and $\|e^{At}\| = 1$ for all $t \in \mathbb{R}$, since $e^{At}\{\xi_k\} = \{e^{i\lambda_k t}\xi_k\}$. The system of eigenvectors of A consists of the unit vectors $\{\delta_{jk}\}_{k=1}^{\infty}$, $j = 1,2,3,\ldots$. It is not complete : its closed linear span is the subspace c_0 of the sequences that converge to zero, and $c_0 \neq c$. Therefore, the representation $t \to e^{At}$ is not a.p. . However, it is scalarly -a.p. . In fact, every functional $f \in B^*$ has the form $f\{\xi_k\} = = \sum \alpha_k \xi_k + \alpha \xi_\infty$, where $\{\alpha_k\}$ is a bounded sequence, α is a scalar, and $\xi_\infty = \lim\limits_{k\to\infty} \xi_k$. Consequently, $f(e^{At}\{\xi_k\}) = \sum \alpha_k \xi_k e^{i\lambda_k t} + \alpha\xi_\infty$ is an a.p.f. .

7°. The a.p. representations of the semigroup $\mathbb{R}_+$ are incorporated in the following general scheme.

LEMMA. *Let* G *be an Abelian topological group, and let* S *be a subsemigroup (with identity) which generates* G, *i.e.,* $G = \{g \mid g = st^{-1},\ s,t \in S\}$. *Then the restriction of the homomorphism* $\beta(S) \to \beta(G)$, *induced by the imbedding* $S \to G$, *to the Sushkevich kernel* K *is an isomorphism of* K *onto* $\beta(G)$.

PROOF. The morphism $\beta(S) \to \beta(G)$ is surjective. In fact, let H denote its image. In view of the commutative diagram

$$\begin{array}{ccc} S & \longrightarrow & G \\ \downarrow & & \downarrow \\ \beta(S) & \longrightarrow & \beta(G) \end{array}$$

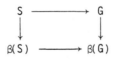

the set HH^{-1} is dense in $\beta(G)$, for $SS^{-1} = G$ and the image of G is dense in $\beta(G)$. But H is a closed subsemigroup of the compact group $\beta(G)$. Hence, it is a group, i.e., $HH^{-1} = H$, and since it is both closed and dense in $\beta(G)$, $H = \beta(G)$. Now suppose that the image of K in $\beta(G)$ is not equal to $\beta(G)$. Then there exists a function $\psi \in C(\beta(G))$, $\psi \neq 0$, which vanishes identically on this image. By pull-back to G, and then to S, ψ becomes an a.p.f. ϕ on S, and as such it can be uniformly approximated by linear combinations of unitary characters of S. Each such character χ defines a unitary character $\overline{\chi}$ of the Bohr compactum $\beta(S)$. If e denotes the identity of the group K, then $\overline{\chi}(t) = \overline{\chi}(te)$ for all $t \in \beta(S)$. Consequently, $\overline{\phi}(te) = \overline{\phi}(t)$ for all $t \in \beta(S)$. Since $te \in K$, $\overline{\phi}(te) = 0$. Hence, $\overline{\phi} = 0$, and so $\phi = 0$ and $\psi = 0$, contrary to the construction of ψ. Thus, the morphism $K \to \beta(G)$ is surjective. It remains to show that it is injective. Let $t_1, t_2 \in K$, $t_1 \neq t_2$. Pick a function $\theta \in C(K)$ such that $\theta(t_1) \neq \theta(t_2)$. Pull-back θ to $\beta(S)$ using the canonical retraction $t \to te$, and then to S. This yields an a.p.f. $\tilde{\theta}$ on S, which can be uniformly approximated by unitary characters of S. But every such character admits a unique extension to a unitary character of G, and hence to a character of the Bohr compactum $\beta(G)$. Linear combinations of characters extend too. Replacing the path $S \to G \to \beta(G)$ in the above diagram by $S \to \beta(S) \to \beta(G)$, we see that the extension just described is isometric. Hence, the fact that $\tilde{\theta}$ is uniformly approximated by linear combinations of unitary characters "extends" to G, and so $\tilde{\theta}$ extends to an a.p. function on G. In the end we obtain a function in $C(\beta(G))$ which separates the points t_1 and t_2.

□

Now, preserving the hypotheses of the Lemma, we consider a.p.f.'s on S. The space $AP(S)$ decomposes under the regular representation as indicated in the Basic Theorem. We thus have the following

THEOREM. *Every a.p.f. ϕ on the subsemigroup S of the Abelian topological group $G = SS^{-1}$ can be written as $\phi = \phi_0 + \phi_1$, where ϕ_0 converges to zero with respect to S and ϕ_1 is a uniform (on S) limit of linear combinations of unitary characters or, equivalently, ϕ_1 extends to an a.p.f. on G.*

□

In the case $S = \mathbb{R}_+$ $(G = \mathbb{R})$ this theorem was obtained by M. Fréchet (1941) in the form : *every a.p.f. on the half-line is the sum of a function that converges to zero and a Bohr function.*

<u>Exercise.</u> *Let A be a compact operator in the reflexive Banach space B such that* $\sup_{t \geq 0} \|e^{At}\| < \infty$. *Then* $B = B_1 \dot{+} B_0$,

where $B_0 = \{x \mid \lim_{t\to+\infty} \|e^{At}x\| = 0\}$ *and* B_1 *is the closed linear span of the eigenvectors of* A *corresponding to purely imaginary eigenvalues.*

8°. An operator A in a Banach space B is said to be *almost periodic* if the representation $k \to A^k$, $k \in \mathbb{Z}_+$, is a.p., i.e., every orbit $\{A^k x\}_{k\geq 0}$ is precompact. If A is a.p., then obviously $\sup_{k\geq 0} \|A^k\| < \infty$, and hence $\rho(A) < 1$. From the Basic Theorem on a.p. representation we derive (taking into account that we are dealing with an Abelian semigroup) the following result.

THE BOUNDARY SPECTRUM SPLITTING-OFF THEOREM. *Let* A *be an a.p. operator in the Banach space* B. *Then* $B = B_1 \dot{+} B_0$, *where* B_0 *and* B_1 *are A-invariant subspaces, defined as follows* :

1) $B_0 = \{x \mid \lim_{x\to\infty} A^k x = 0\}$;

2) B_1 *is the closed linear span of the eigenvectors corresponding to the boundary (or unimodular) spectrum of* A, *i.e., to the points* $\lambda \in \mathrm{spec}_d A$ *lying on the unit circle* $|\lambda| = 1$.

Moreover, the operator $A_1 = A|B_1$ *is invertible, and the group* $\{A_1^k\}_{k\in\mathbb{Z}}$ *that* A_1 *generates is compact in the strong operator topology. If* A *is a contraction, then* A_1 *is isometric and the boundary projection* P *is orthogonal.*

[$\{A_1^k\}_{k\in\mathbb{Z}}$ is obviously the Sushkevich kernel of the representation $k \to A^k$ of $\mathbb{Z}_+$ (to within extension by zero on B_0); it may be referred to as the *Sushkevich kernel of the operator* A.]

□

Remark 1 . In the finite dimensional case all assertion of the theorem become trivial, except for the one concerning contractions. The latter was obtained by the author (1970) in connection with an interesting problem which will be considered at the end of the present chapter.

Remark 2. B_1 *decomposes into a topological direct sum of eigensubspaces (and for contractions the sum is orthogonal).*

Remark 3. The Boundary Spectrum Splitting-Off Theorem admits the following converse : *if* $B = B_1 \dot{+} B_0$, *where* B_0 *and* B_1 *are the subspaces described in the theorem, then* A *is an a.p. operator.* In particular, if $A^k \to 0$ strongly as $k \to \infty$, then A is an a.p. operator. This clearly shows that the restriction of the a.p. A to B_0 has no specific intrinsic structure. For every operator A, $r^{-1}A$ is a.p. for any $r > \rho(A)$. A specific structure may arise when, in addition to the a.p. condition, we impose the constraint $\rho(A) = 1$. Under these conditions, is the existence of nontrivial invariant subspaces guaranteed ? By the Boundary Spectrum Splitting-Off Theorem , this question remains unsettled only in the case where A^k converges strongly to zero.

The Boundary Spectrum Splitting-Off Theorem for a.p. operators proved useful in certain problems of the theory of dynamical systems (M. Yu. Lyubich, 1982). The following corollary of it is particularly useful.

COROLLARY 1 (B. Jamison, 1964). *Let* A *be an a.p. operator which has no boundary (unimodular) eigenvalues* $\lambda \neq 1$. *Then, as* $k \to \infty$, A^k *converges strongly to the projection onto the subspace of fixed vectors of* A.

[Conversely, *if* A^k *converges strongly as* $k \to \infty$, *then* A *is an a.p. operator with no unimodular eigenvalues* $\lambda \neq 1$. This assertion is obvious.]

□

Here the conclusion is stronger than in the ergodic theorem, but at the cost of a stronger hypothesis. Incidentally, the ergodic theorem for a.p. operators is valid without any assumptions on the space B.

COROLLARY 2. *Let* A *be an a.p. operator. Then there exists the strong limit*

$$P_1 = \lim_{n\to\infty} \frac{1}{n} \sum_{k=0}^{n-1} A^k .$$

P_1 *is the projection onto the subspace of fixed vectors of* A, *and it annihilates* B_0 *and all the eigenvectors of* A *correspon-ding to eigenvalues* $\lambda \neq 1$.

In fact, the limit in question vanishes on B_0, and on an arbitrary eigenvector x corresponding to an eigenvalue λ with $|\lambda| = 1$ it vanishes if $\lambda \neq 1$, and it equals x if $\lambda = 1$. Thus, the limit exists on a dense linear manifold, and hence on the whole B, thanks to the boundedness of the powers A^k. That P_1 posses-ses the indicated properties is obvious.

□

COROLLARY 3. *Let* A *be an a.p. operator. Then for every* λ, $|\lambda| = 1$, *there exists the strong limit*

$$P_\lambda = \lim_{n\to\infty} \sum_{k=0}^{n-1} \lambda^{-k} A^k .$$

P_λ *is the projection onto the eigensubspace associated with* λ *(or* 0 *if* $\lambda \notin \mathrm{spec}_d A$*) and annihilates* B_0 *and the other eigen-subspaces.*

□

Thus, the P_λ's are recognized as the projections arising in the general scheme of representation theory. *If* A *is a con-traction, than all* P_λ *are orthogonal projections.*

Exercise 1. *Suppose that the operator* A *possesses a complete system of eigenvectors associated to eigenvalues lying on the unit circle and that* $\sup_{k\geq 0} \|A^k\| < \infty$. *Then* A *is invertible and the re-presentation* $k \to A^k$ *of* $\mathbb{Z}$ *is a.p.* (in this sense A is two-sided a.p.)

Exercise 2. *An invertible operator* A *is two-sided a.p. if and only if it is a.p. and* $\sup_{k<0} \|A^k\| < \infty$.

The simplest interesting example of an a.p. operator is a compact operator A with the property that $\sup_{k\geq 0} \|A^k\| < \infty$. If $\rho(A) = 1$, then $\dim B_1 < \infty$ and $\rho(A|B_0) < 1$. As it turns out, these properties are preserved under "small" perturbations (K.

Yosida - S. Kakutani, 1941). The proof of this fact given below rests on a reduction to the a.p. case.

THEOREM. *Let* $A = U + V$, *where* U *is a compact operator,* $\rho(V) < 1$, *and* $\sup_{k\geq 0} \|A^k\| < \infty$. *Then* A *is an a.p. operator,* $\dim B_1 < \infty$, *and* $\rho(A|B_0) < 1$.

PROOF. With no loss of generality we may assume that $\|A\| \leqslant 1$. Our starting point is the formula

$$A^n = \sum_{k=0}^{n-1} V^k U A^{n-k-1} + V^n, \quad n = 1,2,3,\dots, \tag{1}$$

which is readily verified by induction. The sum of the terms involving the compact factor U is compact, and $\rho(V^n) < 1$. Hence, the assumptions of the theorem remain valid on replacing A by A^n. Since $\|V^n\| \to 0$ as $n \to \infty$, we can arrange that $\|V^n\| < 1$. On the other hand, if the conclusion of the theorem is valid for A^n, then it is valid for A. Therefore, we can assume from the very beginning that $\|V\| < 1$. Then it follows from (1) that the orbit $O(x)$ of any vector x can be arbitrarily well approximated by finite sums of precompact sets $V^k UO(x)$. Consequently, $O(x)$ is precompact, i.e. , A is an a.p. operator.

We next put $A_1 = A|B_1$, $U_1 = PU|B_1$, and $V_1 = PV|B_1$, where P is the boundary projection. Then $A_1 = U_1 + V_1$, where U_1 and V_1 are operators in B_1, U_1 is compact, and $\|V_1\| < 1$ ($\|P\| \leqslant 1$ because $\|A\| \leqslant 1$). Since A_1 is an invertible isometry, $U_1 = A_1(E - A_1^{-1}V_1)$ and $\|A_1^{-1}V_1\| < 1$. Therefore, U_1 is invertible, and so in B_1 there exists a compact invertible operator. This forces $\dim B_1 < \infty$.

Replacing A, if necessary, by one of its powers (which does not affect the projection P), we may assume that $\|(E - P)V\| < 1$. Set $A_0 = A|B_0$, $U_0 = (E - P)U|B_0$, and $V_0 = (E - P)V|B_0$. Then $A_0 = U_0 + V_0$, where U_0 and V_0 act in B_0, U_0 is compact, and $\|V_0\| < 1$. Moreover, $\|A_0\| \leqslant 1$ and A_0 has no eigenvalues on the unit circle. We show that $\rho(A_0) < 1$. Suppose $\rho(A_0) = 1$.

Then for some λ, $|\lambda| = 1$, there is a sequence $\{x_k\} \subset B_0$, $\|x_k\| = 1$, such that $A_0 x_k - \lambda x_k \to 0$ as $k \to \infty$. This gives $U_0 x_k + (V_0 - \lambda E)x_k \to 0$ and $R_\lambda U_0 x_k + x_k \to 0$, where R_λ denotes the resolvent of V_0. Since U_0 is compact, we can extract from $\{U_0 x_k\}$, and hence from $\{x_k\}$, a subsequence which converges to a vector x, $\|x\| = 1$. But then $A_0 x - \lambda x = 0$: contradiction.

□

COROLLARY. *Suppose there is an* $m \in \mathbb{Z}_+$ *such that* $A^m = U + V$, *where* U *is a compact operator and* $\|V\| < 1$. *If* $\sup_{k\geq 0} \|A^k\| < \infty$, *then* A *is an a.p. operator,* $\dim B_1 < \infty$, *and* $\rho(A|B_0) < 1$.

□

Under these assumptions the mean $n^{-1}\sum_{k=0}^{n-1} A^k$ converges in the operator norm, i.e., we may say that the *uniform ergodic theorem* holds (K. Yosida - S. Kakutani, 1941).

4. NONNEGATIVE A.P. REPRESENTATIONS

1°. Let B be a real Banach space in which there is given a closed convex cone B_+ (for brevity we call B a *space with a cone* B_+). [Recall that a subset of a real vector space is called a *cone* if it is invariant under multiplication by scalars $\alpha \geq 0$ and does not contain simultaneously x and $-x$ if $x \neq 0$. A cone is convex if and only if it is invariant under addition, i.e., it is an additive semigroup. Hereafter by "cone" we shall mean closed convex cone.] The vectors $x \in B_+$ are termed *nonnegative*. We can thus define an order on B, setting $x \geq y$ whenever $x - y \in B_+$. Thus, $x \geq 0 \Leftrightarrow x \in B_+$. This order is compatible with the linear operations and with passage to limit in B. The cone B_+ is called *solid* if its interior $\operatorname{int} B_+$ is not empty. The vectors $x \in \operatorname{int} B_+$ are termed *positive* (in which case we write $x > 0$). A Banach space with a solid cone is called a *Krein space*. An important example of Krein space is $C(Q)$, the space of continuous real-valued functions on the compact topological space Q, with the usual norm and the cone $C_+(Q)$ of pointwise-nonnegative functions.

We say that the cone B_+ is *reproducing* if every vector $x \in B$ can be written as $x = x' - x''$, with $x', x'' \geqslant 0$.

Exercise. *Every solid cone is reproducing.*

Example. Consider the real space $L_p(S;ds)$, $1 \leqslant p < \infty$, over an arbitrary space S with a measure ds. In $L_p(S;ds)$ the cone of all functions which are positive a.e. is reproducing, but not necessarily solid. The same is true for the cone of (nonnegative) measures in the space of all finite real-valued measures.

The cone B_+ is called *total* if every vector $x \in B$ can be written as $x = \lim_{k\to\infty} (x_k' - x_k'')$, with $x_k' \geqslant 0$, $x_k'' \geqslant 0$.

Let B be a space with a cone B_+. Then B^* is a space with the cone $B_+^* = \{f \mid f(x) \geqslant 0, \forall x \in B_+\}$. For $f \in B^*$ we say that f is *positive*, and write $f > 0$ if $f(x) > 0$ for all $x \in B_+$, $x \neq 0$ [this terminology does not mean that $f \in \text{int } B_+^*$; it is meaningful even if the cone B_+^* is not solid].

Exercise. *If* $x \geqslant 0$, $x \neq 0$, *then there exists a linear functional* $f \geqslant 0$ *such that* $f(x) > 0$.

We next define a cone that arises naturally in $L(B)$. It consists of the operators A with the property that $AB_+ \subset B_+$. We say that A is *positive*, and write $A > 0$, if Ax int B_+ for all $x \geqslant 0$, $x \neq 0$. The operators $A \geqslant 0$ form a semigroup (with identity) with respect to multiplication.

The general form of a nonnegative operator in $C(Q)$ is : $(A\phi)(u) = \int_Q \phi \, d\alpha_u$, where $\{\alpha_u\}_{u\in Q}$ is a family of measures on Q. The set $\overline{\bigcup_u \operatorname{supp} \alpha_u}$ is called the *support* of the operator A and is denoted by supp A.

Exercise 1. *A point* $v \in Q$ *belongs to* supp A *if and only if* $A\phi \neq 0$ *for every function* $\phi \geqslant 0$ *such that* $\phi(v) > 0$.

<u>Exercise 2.</u> *If* $A \geqslant 0$, *then* $\|A\| = \|A\mathbb{1}\|$ *(in* $C(Q)$*).*

An operator A in $C(Q)$ is called a *stochastic* or *Markov operator* if $A \geqslant 0$ and $A\mathbb{1} = \mathbb{1}$. In this case all α_u's are probability measures.

<u>Example.</u> Let $F : Q \to Q$ be a continuous map. Then the operator A in $C(Q)$ defined by : $(A\phi)(u) = \phi(F(u))$ is stochastic.

2°. We say that the representation T of the topological semigroup S in a space with a cone is *nonnegative*, and we write $T \geqslant 0$, if $T(s) \geqslant 0$ for all s S. A representation T in $C(Q)$ is called *stochastic* if $T(s)$ is a stochastic operator for all s. The set of all stochastic operators in $C(Q)$ is a semigroup (with identity).

In 1907 O. Perron, and after him G. Frobenius developed the important and beautiful spectral theory of nonnegative matrices (i.e., here, matrices with nonnegative entries ; a detailed exposition of this theory can be found in the monograph of F. R. Gantmakher [14]). It can be regarded as the theory of nonnegative operators (or nonnegative representations of the semigroup $\mathbb{Z}_+$) in the space $C(Q)$ over a finite set Q. A central result in the Perron-Frobenius theory is the theorem asserting that *the spectral radius of a nonnegative matrix is one of its eigenvalues, and to it there corresponds a nonnegative eigenvector and a nonnegative linear functional dual to this vector* (this result and its generalizations are often referred to as the *Perron-Frobenius Theorem*).

In 1937 M. A. Rutman generalized this result to nonnegative compact operators in a space with a solid cone. Soon after M. G. Krein obtained a number of further generalizations that did not require that the operators be compact. A systematic exposition of these results was given by their authors in 1948. Since then they became widely known and found numerous applications.

In 1964 M. Rosenblatt showed that under certain assumptions the Perron-Frobenius Theorem (as well as other facts) carries over to nonnegative a.p. operators in the space $C(Q)$ over an arbitrary compact topological space Q. As it turns out, the theorem is valid even for nonnegative a.p. representations in a space with a cone under rather broad assumptions. This generalization, recently obtained by M. Yu. Lyubich and the author, can be stated as follows:

THEOREM. *Let* B *be a Banach space with a total cone,* S *a topological semigroup, and* T *a nonnegative a.p. representation of type* 1×1 *(recall that this last condition is automatically fulfilled in the Abelian case). Suppose that there is a vector* $x_1 \in B$ *whose orbit is separated from zero* : $\inf_s \|T(s)x_1\| > 0$. *Then there exists a fixed (invariant) vector* $h \geqslant 0$ *(i.e.,* $T(s)h = h$ *for all* $s \in S$*) and an invariant linear functional* $\mu \geqslant 0$ *dual to* h *(i.e.,* $T^*(s)\mu = \mu$ *for all* $s \in S$, *and* $\mu(h) > 0$*).*

PROOF. The boundary projection P is the unit element of the Sushkevich kernel K of representation T. Therefore, $P \geqslant 0$, and in fact $P \neq 0$ (otherwise the closure of the orbit of any vector would contain the null vector). Since the cone B_+ is total, $P|B_+ \neq 0$. Now set $h = \int (Ax_+)dA$, where dA is the normalized Haar measure on K and $x_+ \in B_+$ is an arbitrary vector such that $Px_+ \neq 0$. Then $h \neq 0$ (and obviously $h \geqslant 0$). To show this, it suffices to pick a $\nu \in B_+^*$ such that $\nu(Px_+) > 0$. Then $\nu(h) = \int \nu(Ax_+)dA > 0$, since the integrand is nonnegative and different from zero at the point P (it is also continuous, K being endowed with the strong operator topology). Next, K, as a two-sided ideal in the Bohr compactum, is invariant under all operators $T(s)$. Consequently, $T(s)h \equiv h$.

The required functional μ is defined by

$$\mu(x) = \int \nu(Ax)dA.$$

Then $\mu \geqslant 0$, μ is invariant, and

$$\mu(h) = \int \nu(Ah)dA = \nu(h) \int dA = \nu(h) > 0 .$$

□

Remark 1. *If* x_1 *is a weight vector corresponding to a unitary character, then its orbit is separated from zero.*

Remark 2. The Perron-Frobenius Theorem is covered by the theorem just proved (as is Rutman's Theorem) if one requires also that the powers of the matrix (respectively, operator) be bounded. The existence of the vector x_1 follows from the general theory of compact operators.

COROLLARY. *Let* T *be a nonnegative a.p. representation of type* 1×1 *in* $C(Q)$. *Suppose the spectral radius* $\rho(T(s)) = 1$ *for all* s. *Then there exists an invariant function* $h \geq 0$ *and an invariant measure* μ *such that* $\mu(h) > 0$.

In fact, in this case the orbit of the constant function $\mathbb{1}$ is separated from zero :

$$\|T(s)\, \mathbb{1}\| = \|T(s)\| \geq \rho(T(s)) = 1$$

□

3°. A matrix $A = (a_{ik})_{i,k=1}^{n}$ is said to be *indecomposable (according to Frobenius)* if for every pair j,ℓ, $1 \leq j,\ell \leq n$, there is a positive integer m such that $a_{j\ell}^{(m)} > 0$, where $a_{ik}^{(m)}$ denotes the (i,k)-entry in the matrix A^m. Nonnegative indecomposable matrices enjoy a number of remarkable spectral properties. In particular, *the boundary (unimodular) spectrum of such a matrix* A *is (provided* $\rho(A) = 1$*) a group, and hence the group of roots of unity of a certain degree.*

[This implies that *the boundary spectrum of every nonnegative matrix* A *with* $\rho(A) = 1$ *consists of roots of unity.* N. N. Bogolyubov and S. G. Krein (1947) established this property of the boundary spectrum for nonnegative compact operators in $C(Q)$ and even in a wider class of Krein spaces.]

Moreover, *the full spectrum of* A *is invariant under the action of this group. The eigensubspaces associated with the boundary spectrum are one-dimensional.*

In the classical theory these facts seem somewhat misterious. However, they can be incorporated in a broader picture, where they admit a transparent interpretation. We begin with general definitions and remarks of geometric nature.

Let T be a nonnegative (not necessarily a.p.) representation of the semigroup S in the space B with a total cone B_+. We say that T is F-*indecomposable* if for every $x \geqslant 0$, $x \neq 0$, and every $f \geqslant 0$, $f \neq 0$, there is an $s \in S$ such that $f(T(s)x) > 0$. [We use the letter F to avoid confusion with indecomposability in the sense of representation theory, and also to indicate that this notion is a descendant of the notion of a Frobenius-indecomposable matrix.] If B is a Krein space and $T \geqslant 0$ is an F-indecomposable representation, we say that T is *primitive* if for every $x \geqslant 0$, $x \neq 0$, there is an $s \in S$ such that $T(s)x > 0$ (in the opposite case T is called *imprimitive*). These notion carry over to an individual operator A via the representation $k \to A^k$ of $\mathbb{Z}_+$. For matrices this reduces to the classical definition, whose semigroup nature is manifest.

Exercise 1. *If* $A > 0$, *then* A *is primitive.*

Exercise 2. *Let* $\mu \geqslant 0$, $\mu \neq 0$, *be an invariant linear functional of an* F-*indecomposable representation. Then* $\mu > 0$.

THEOREM. *Suppose the* F-*indecomposable representation* T *in a Krein space possesses a fixed vector* $h \geqslant 0$, $h \neq 0$. *Then* $h > 0$.

PROOF. Let $f \geqslant 0$, $f \neq 0$, and let $s \in S$ be such that $f(T(s)h) > 0$, i.e., $f(h) > 0$. Then $h > 0$ by the separating hyperplane theorem.

□

COROLLARY. *Under the assumptions of the theorem the subspace of* T-*fixed vectors is one-dimensional.*

PROOF. Let $h > 0$ and $x \neq 0$ be fixed vectors. Consider the line $h + \tau x$ $(\tau \in \mathbb{R})$. It is not contained in the cone, but for small values of $|\tau|$ its points lie inside the cone. It follows that it intersects the boundary of the cone for some τ_0. By the theorem, this can happen only if $h + \tau_0 x = 0$. □

Exercise. *For a projection* $P \geq 0$ *in a Krein space* F-*indecomposability implies positivity (and hence primitivity). The general form of such a projection is* $Px = \mu(x)h$, *where* $h > 0$, $\mu > 0$, *and* $\mu(h) = 1$. *Therefore, every* F-*indecomposable projection in a Krein space is one-dimensional.*

THEOREM. *Suppose the representation* $T \geq 0$ *in the Krein space* B *is primitive and possesses a fixed vector* $h \geq 0$, $h \neq 0$. *Then there are no finite dimensional* T-*invariant subspaces, other than* Lin(h), *in which the orbit of every vector* $x \neq 0$ *is separated from zero and infinity.*

PROOF. Let L be such an invariant subspace. We may assume that $h \in L$. Since $h > 0$, the cone $L_+ = B_+ \cap L$ is solid in L. The representation $T|L$ remains primitive, but in addition it is a.p., thanks to its boundedness. The Bohr compactum of $T|L$ is a group, since the orbits are separated from zero. The inverses of the operators $T(s)|L$ exist and are nonnegative. But then each $T(s)|L$ maps the extreme points of L_+ again into such points, which contradicts the primitivity of $T|L$ if $L \neq \text{Lin}(h)$.

□

COROLLARY 1. *Suppose the representation* $T \geq 0$ *in the Krein space* B *is a.p. and possesses a fixed vector* $h > 0$. *Then* T *is primitive if and only the boundary subspace* B_1 *is one-dimensional.*

PROOF. The necessity of this condition is a straightforward consequence of the theorem. To prove its sufficiency, we use the fact that the boundary projection P reduces to $P = \mu(\cdot)h$, where $\mu \geq 0$ $(\mu(h) = 1)$ is an invariant linear functional. Now let $x \geq 0$, $x \neq 0$. Consider the vector function $T(s)x - \mu(x)h = T(s)(x - Px)$. Since $x - Px \in B_0$, given any $\varepsilon > 0$ one can find an s such that $\|T(s)x - \mu(x)h\| < \varepsilon$. But $\mu(x)h > 0$. Taking ε small enough, we conclude that $T(s)x > 0$.

□

COROLLARY 2. *Let* T *be a primitive a.p. representation of the Abelian semigroup* S *in the Krein space* B. *Suppose* T *possesses a vector whose orbit is separated from zero. Then there exists an invariant vector* $h > 0$ *and an invariant linear functional* $\mu > 0$, $\mu(h) = 1$, *such that* $\lim_s \|T(s)x - \mu(x)h\| = 0$ *for all* $x \in B$.

□

4°. We next turn to F-indecomposable nonnegative a.p. representations of type *1 × 1* in $C(Q)$. The theory of this class of representations, an exposition of which is given below, has been developed by M. Yu. Lyubich jointly with the author. We shall assume that $\rho(T(s)) = 1$ for all $s \in S$ (this and the properties of the representation listed above will be not mentioned explicitly in what follows). Then the general theorem guarantee the existence of an invariant function $h > 0$ and an invariant measure $\mu > 0$ satisfying $\mu(h) = 1$.

LEMMA 1. T *is equivalent to a stochastic representation. The equivalence is realized by the operator* $\hat{h}$ *of multiplication by the function* h.

PROOF. Set $T(s) = \hat{h}^{-1}T(s)\hat{h}$. Then obviously $T(s) \geqslant 0$ and $T(s)\mathbb{1} = h^{-1}T(s)h = \mathbb{1}$, since $T(s)h = h$.

□

In the following we shall assume (in arguments, but not in the statement of the final results) that the representation T is stochastic. Then all operators belonging to the Bohr compactum of T, including the boundary projection P, are stochastic.

<u>Exercise.</u> $\operatorname{Im} P$ *is a Krein space (with the cone* $\operatorname{Im} P \cap B_+$*).*

LEMMA 2. $\operatorname{supp} P = Q$.

PROOF. Suppose $\operatorname{supp} P \neq Q$. Then there exists a function $\phi \geqslant 0$, $\phi \neq 0$, such that $\phi|\operatorname{supp} P = 0$ (throughout this section we consider only continuous functions). Hence, $P\phi = 0$, and since $\operatorname{Ker} P$ is T-invariant, we have $PT(s)\phi = 0$. Since $T(s)\phi \geqslant 0$, this gives $T(s)\phi|\operatorname{supp} P = 0$ for all s, which contradicts the F-indecomposability of representation T.

□

The next step is to identify in Q the points that are not separated by functions belonging to $\operatorname{Im} P$. This yields a compact quotient space $\tilde{Q}$ together with a canonical homomorphism $i : \operatorname{Im} P \to C(\tilde{Q})$ (the points of $\tilde{Q}$ are called *imprimitivity classes*). It is clear that i preserves the norm as well as the order; moreover, $i\phi \geqslant 0 \Leftrightarrow \phi \geqslant 0$.

LEMMA 3. $\operatorname{Im} i = C(Q)$. *Thus,* i *is both an isometry and an order isomorphism of the Krein spaces* $\operatorname{Im} P$ *and* $C(\tilde{Q})$.

PROOF. By a well-known theorem of Stone (which is the "order" analogue of the Stone-Weierstrass Theorem; see, for example, the monograph of M. Day [11]), it suffices to show that if $\psi_1, \psi_2 \in \operatorname{Im} i$, then the function ψ, $\psi(\xi) = \max(\psi_1(\xi), \psi_2(\xi))$, also belongs to $\operatorname{Im} i$. Let $\psi_1 = iP\phi_1$ and $\psi_2 = iP\phi_2$. Set $\theta(u) = \max(P\phi_1(u), P\phi_2(u))$. Then $\theta \geqslant P\phi_1$, whence $P\theta \geqslant P\phi_1$. Similarly, $P\theta \geqslant P\phi_2$. Thus, $P\theta \geqslant \theta$. Consider the difference $\omega = P\theta - \theta \geqslant 0$. Since $P\omega = 0$, we have $\omega|\operatorname{supp} P = 0$, that is, by Lemma 2, $\omega = 0$. Observing that the homomorphism i commutes with the operation of taking the pointwise maximum, we conclude that $i = \max(iP\phi_1, iP\phi_2) = \max(\psi_1, \psi_2) = \psi$, i.e., $\psi \in \operatorname{Im} i$.

□

Now let A be an arbitrary operator belonging to the Sushkevich kernel K of representation T. Put $\tilde{A} = i(A|\operatorname{Im} P)i^{-1}$. The operator $\tilde{A}$ in $C(\tilde{Q})$ is stochastic. It is invertible, and its inverse is also stochastic, since K is a group, with identity element P, consisting entirely of stochastic operators. Consequently, $(\tilde{A}\psi)(\xi) = \psi(a^{-1}\xi)$ for all $\psi \in C(\tilde{Q})$, where $a : \tilde{Q} \to \tilde{Q}$ is a homeomorphism uniquely determined by the operator $\tilde{A}$. In fact, $\tilde{A}^*$, acting in the space of measures on the compact space Q, maps the simplex of probability measures bijectively onto itself. In particular, $\tilde{A}^*\delta_\xi = \delta_{a^{-1}\xi}$, where δ_ξ denotes the Dirac measure, i.e., the unit-mass measure supported at the point ξ. Its image under $\tilde{A}^*$ is again a Dirac measure, since $\tilde{A}^*$ maps the set of extreme points of the simplex bijectively onto itself, and these points are precisely the Dirac measures. The map a is defined by the indicated equality, and is a homeomorphism, as is readily verified. Regarding a as a function $A \to a(A)$ from K into Homeo $\tilde{Q}$, we obtain an action of K on $\tilde{Q}$. In particular, setting $A = A(s) \equiv T(s)P$, we have $\tilde{A}(s) = i(T(s)|\mathrm{Im}\ P)i^{-1}$ and $(\tilde{A}(s)\psi)(\xi) = \psi(a(s)^{-1}\xi)$, where $a(s) = a(A(s))$.

LEMMA 4. *The action* a *is transitive, i.e.,* $\tilde{Q}$ *is a homogeneous space of the group* K.

PROOF. Suppose this is not the case. Let $\xi \in \tilde{Q}$. Then the orbit $O(\xi) \neq \tilde{Q}$. On the other hand, $O(\xi)$ is compact. Choose a function $\psi \in C_+(\tilde{Q})$, $\psi \neq 0$, such that $\psi|O(\xi) \equiv 0$. By Lemma 3, $\psi = i\phi$, where $\phi \in \mathrm{Im}\ P$, $\phi \geqslant 0$, $\phi \neq 0$. If $u \in Q$ is any point in the class ξ, then $(T(s)\phi)(u) = \psi(a(s)^{-1}\xi) = 0$ for all $s \in S$. This contradicts the F-indecomposability of representation T.

□

We thus obtained a *Structure Theorem* for representations of the class under consideration. Its complete formulation goes as follows.

THEOREM. *Let* S *be a topological semigroup and let* Q *be a compact space. Let* T *be a nonnegative* F*-indecomposable a.p. representation of type* 1×1 *of* S *in the space* $C(Q)$ *such that* $\rho(T(s)) = 1$ *for all* $s \in S$. *Let* P *be the identity element of the Sushkevich kernel* K *of* T, h *an invariant function,* $h > 0$, $\hat{h}$ *the operator of multiplication by* h, $\tilde{Q}$ *the compact space obtained by identifying in* Q *the points that are not separated by functions in* $\mathrm{Im}(\hat{h}^{-1}P\hat{h})$, *and* $i : \mathrm{Im}\ P \to C(\tilde{Q})$ *the canonical homomorphism.*

Then $i\hat{h}^{-1}$ *is an isomorphism of the boundary subspace* $\mathrm{Im}\ P$ *onto* $C(\tilde{Q})$ *which intertwines* $T|\mathrm{Im}\ P$ *and the representation generated by the transitive action of the compact group* K *on* $\tilde{Q}$ *(to within the canonical homomorphism* $S \to K$*).*

□

In formulating below the spectral consequences of this theorem (which, needless to say, refer to the complexified representation) we shall assume that all its hypotheses are in force, and in the proofs we shall assume that T is generated by a transitive action of the compact group K on the compact space Q.

COROLLARY 1. *The weight subspaces of* T *associated with unitary weights are one-dimensional.*

PROOF. Let χ be a unitary weight and ϕ an associated weight function. Then $\phi(a(g)^{-1}u) = \chi(g)\phi(u)$ for all u Q and all $g \in K$. Thanks to the transitivity of a, this identity defines, for fixed u, the function ϕ on the whole space Q.

□

Remark. It follows from the same identity that *in the case of the transitive action of a group, and hence in the stochastic case isometric to it, every weight function has constant modulus. In the general case the modulus is proportional to* h *(for all unitary weights).* We mention also that *in the case of the action of a group on itself the weight subspaces are spanned by the corresponding weights.*

COROLLARY 2. *The unitary weights of representation* T *form a group.*

PROOF. Let ψ_1 and ψ_2 be weight functions associated with the unitary weights χ_1 and χ_2. Then the function $\psi_1\psi_2^{-1}$ is associated with the weight $\chi_1\chi_2^{-1}$.

□

This assertion can be sharpened as follows.

THEOREM. *Let* χ *be a unitary weight of representation* T. *Then the representation* $s \to \chi(s)T(s)$ *is equivalent to* T *(and if* T *is stochastic the equivalence is realized by the operator* $\hat{\psi}$ *of multiplication by the associated weight function* ψ*).*

PROOF. If T is stochastic, we may assume that $|\psi| = \mathbb{1}$. Representation T has the form

$$(T(s)\phi)(u) = \int_Q \phi \, d\tau_{s,u} ,$$

where $\tau_{s,u}$ are probability measures. Consequently,

$$\int \psi \, d\tau_{s,u} = \chi(s)\psi(u) .$$

Since $|\chi(s)\psi(u)| \equiv 1$, $\psi|\text{supp } \tau_{s,u} = \chi(s)\psi(u)$. Therefore, for every function θ,

$$(T(s)(\psi\theta))(u) = \int \psi\theta d\tau_{s,u} = \chi(s)\psi(u) \int \theta \, d\tau_{s,u} =$$

$$= \chi(s)\psi(u)(T(s)\theta)(u),$$

i.e., $\hat{\psi}^{-1}T(s)\hat{\psi} = \chi(s)T(s)$.

□

We remark that *the primitive representations are characterized within the class under consideration by the property that the compact space* $\tilde{Q}$ *reduces to one point.* In fact, this is equivalent to $\dim C(\tilde{Q}) = 1$, i.e., to the boundary subspace being one-dimensional.

In the Abelian case the picture described above becomes even more transparent.

THEOREM. *Suppose the semigroup* S *is Abelian. Then the compact space* $\tilde{Q}$ *is homeomorphic to the Sushkevich kernel* K *of representation* T, *and* $T|\text{Im } P$ *is equivalent to the regular representation of the compact group* K *in* $C(K)$ *(up to the restriction of the latter to a dense semigroup).*

PROOF. It suffices to show that the action of K on $\tilde{Q}$ is effective. Suppose $a(A_0)\xi_0 = \xi_0$ for some $A_0 \in K$ and $\xi_0 \in \tilde{Q}$. Then the set of all fixed points of the homeomorphism $a(A_0)$ is nonempty and invariant under all $a(A)$ with $A \in K$ (the latter thanks to the fact that S is Abelian). Since the action is transitive, this set is equal to $\tilde{Q}$, i.e., $a(A_0) = \text{id}$.

□

COROLLARY. *In the Abelian case the group of unitary weights of the given representation is isomorphic to the dual group* K^*.

□

Let us review the theory developed to this point on the case of an individual a.p. operator $A \geq 0$ in $C(Q)$. This case has been studied (under the F-indecomposability and stochasticity assumptions) in works of M. Rosenblatt (1964), B. Jamison (1964), and B. Jamison and R. Sine (1969) in connection with applications to probability theory. A considerable part of the results described below belongs to these authors.

THEOREM. *Let* $A \geq 0$ *be an a.p. operator in* $C(Q)$ *such that* $\rho(A) = 1$. *Then* :

1. There exist an invariant function $h \geq 0$ *and an invariant measure* $\mu > 0$ *(with* $\mu(h) = 1$*).*

2. If A *is F-indecomposable, then* :

a) $h > 0$, $\mu > 0$;

b) the eigenvalues lying on the unit circle form a group ;

c) the corresponding eigensubspaces are one-dimensional ;

d) if $\lambda \in \mathrm{spec}_d A$ *and* $|\lambda| = 1$, *then the operator* λA *is similar to* A, *and so the full spectrum of* A *is invariant under multiplication by* λ ;

e) A *is similar to a stochastic operator via the operator of multiplication by* h.

3. Suppose A *is F-indecomposable and stochastic. Let* P *be the boundary projection of* A, *and let* $\tilde{Q}$ *denote the compact space obtained by identifying in* Q *the points that are not separated by the functions in* $\mathrm{Im}\, P$. *Then* $\tilde{Q}$ *can be turned into a compact Abelian group* K *(isomorphic to the Sushkevich kernel of* A*) such that* $A|\mathrm{Im}\, P$ *becomes the operator induced in* $C(K)$ *by a topologically transitive translation on* K. [The existence of such a translation means that K is a *monothetic group*, i.e., it contains a dense cyclic subgroup $\{g^n\}_{n \in \mathbb{Z}}$. The element g is called a *topological generator of* K.]

4. The boundary spectrum of A, *regarded as a group, is isomorphic to the dual group* K^*.

5. *An F-indecomposable operator* A *is primitive if and only if the quotient space* $\tilde{Q}$ *reduces to one point.*

6. *If* A *is primitive, then* $\lim_{k\to\infty} \|A^k\phi - \mu(\phi)h\| = 0$ *for all* $\phi \in C(Q)$.

□

Assertion 4 has the following interesting consequence.

COROLLARY (B. Jamison - R. Sine, 1969). *Let* $A \geq 0$ *be an F-indecomposable a.p. operator in* $C(Q)$ *such that* $\rho(A) = 1$. *If the compact space* Q *is connected, then the boundary (unimodular) spectrum of* A *contains no roots of unity different from* 1.

PROOF. We may assume that A is stochastic. The canonical map $Q \to \tilde{Q}$ is continuous. Hence, the group $K \equiv \tilde{Q}$ is connected, and so K^* contains no elements of finite order different from the identity.

□

5°. A stochastic representation T in $C(Q)$ is necessarily contractive : $\|T(s)\| = \|T(s)\mathbb{1}\| = \|\mathbb{1}\| = 1$. Moreover, it possesses a unit weight. It turns out that every contractive representation which has a unit weight and is, in a certain sense, indecomposable, is equivalent to an F-indecomposable stochastic representation. As such, it enjoys all the spectral properties of the latter. However, to prove the equivalence theorem we need to establish first a number of properties of the given representation.

Let T be a representation of the topological semigroup in the complex space $C(Q)$. Set

$$(T(s)\phi)(u) = \int_Q \phi \, d\tau_{s,u}, \quad u \in Q,$$

where $\tau_{s,u}$ are complex measures. Obviously, $\|T(s)\| = \sup \|\tau_{s,u}\|$. Hence, the contractivness condition for T is written : $\|\tau_{s,u}\| \leq 1$ for all $s \in S$ and all $u \in Q$. The representation T is called F-*indecomposable* if $\bigcup_s \operatorname{supp} \tau_{s,u} = Q$ for all u. The F-indecomposability of an operator A in $C(Q)$ is defined accordingly. For a nonnegative representation (operator) this definition is equivalent to the previous one.

LEMMA. *Let* χ *be a unitary weight of the* F-*indecomposable contractive representation* T. *Then every weight function* ψ *associated with* χ *has constant modulus.*

PROOF. Suppose that $||\psi|| = 1$ and $|\psi(u_0)| = 1$. If $|\psi(u_1)| < 1$ for some point u_1, then in view of the F-indecomposability of T we may assume that $u_1 \in \text{supp}\ \tau_{s_0,u_0}$ for a suitable s_0. But then it follows from the equality $\chi(s)\psi(u_0) = \int \psi\, d\tau_{s_0,u_0}$ that $|\psi(u_0)| < 1$: contradiction.

□

COROLLARY 1. *The weight subspaces associated with the unitary weights of an* F-*indecomposable contractive representation are one-dimensional.*

□

COROLLARY 2. *Suppose the* F-*indecomposable contractive representation* T *possesses at least one unitary weight. Then* $||\tau_{s,u}|| = 1$ for all s,u.

In fact, if $||\tau_{s,u}|| = 1$ for some s and u, then $|\psi(u)| < 1$ for a weight function ψ such that $||\psi|| = 1$.

□

In the case of a finite Q, Corollary 2 becomes a theorem of O. Taussky (1949) asserting that if the matrix $A = (a_{jk})_{j,k=1}^{n}$ is indecomposable, $\sum_k |a_{jk}| \leqslant 1$ for all j, and the inequality is strict for at least one j, then $\rho(A) < 1$.

Exercise. *The general form of an* F-*indecomposable orthogonal projection is* $Px = \mu(x)h$, *where* μ *is a complex measure,* $\text{supp}\ \mu = Q$, $||\mu|| = 1$, $\mu(h) = 1$, *and* $|h| = 1$.

THEOREM. *Let* χ *be a unitary weight of the* F-*indecomposable contractive representation* T. *Then the representation* $T_\chi(s) = \chi(s)T(s)$ *is equivalent to an* F-*indecomposable stochastic repre-*

sentation (and the equivalence is realized by the operator $\hat{\psi}$ *of multiplication by the weight function* ψ *associated with* χ*).*

PROOF. We shall assume that $\|\psi\| = 1$. The function ψ is invariant under T_χ. Hence, the constant function $\mathbb{1}$ is invariant under the representation $\hat{\psi}^{-1}T_\chi\hat{\psi}$. Since $\hat{\psi}^{-1}T_\chi\hat{\psi}$ is contractive, it is also nonnegative (if the complex measure ν is such that $\|\nu\| \leqslant 1$ and $\nu(\mathbb{1}) = 1$, then $\nu \geqslant 0$). We thus conclude that $\hat{\psi}^{-1}T_\chi\hat{\psi}$ is stochastic. It is obviously F-indecomposable together with T.

□

In particular, if $\chi = \mathbb{1}$, then T itself is equivalent to a stochastic (and F-indecomposable) representation.

The theorem just proved has a conditional character in the sense that it is assumed that a unitary weight exists. However, if one requires that T be a.p. and that there should exist a function whose orbit is separated from zero, then in the case of an Abelian semigroup S the existence of a unitary weight χ is guaranteed by the Basic Theorem. In this case the set of all unitary weights turns out to be a coset of S^* modulo the subgroup K^*, where K is the Sushkevich kernel of representation T. For a finite set Q this coset can be described explicitly in terms of the so-called graph of a matrix (Yu. I. Lyubich and M. I. Tabachnikov, 1969).

6°. If $|Q| < \infty$, *then for every contraction* A ($\rho(A) = 1$) *in the real space* C(Q) *all the points of the boundary (unimodular) spectrum are roots of unity.* For an F-indecomposable contraction this follows from the group property, and for an arbitrary contraction - from the fact that one can partition Q in such a manner that the diagonal blocks of the matrix that correspond to the elements of the partition are F-indecomposable.

THEOREM (Yu. I. Lyubich, 1970). *Let* B *be a real finite dimensional vector space. Let* $A : B \to B$ *be a contraction such that* $\rho(A) = 1$. *Then all the points in the boundary spectrum of* A *are roots of unity if and only if* B *contains no orthocomplemented two-dimensional Euclidean subspaces.*

Therefore, *the indicated property of the boundary spectrum is fulfilled in every space in which the balls are the polyhedra.*

PROOF. NECESSITY. This part is almost obvious. Let L be a two-dimensional subspace admitting an orthogonal complement, and let P denote the projection onto L. Let $U(\alpha)$ be the rotation in L by an angle α incommensurable with 2π. Then the boundary spectrum of the contraction $U(\alpha)P$ contains the point $\lambda = e^{i\alpha}$, which is not a root of unity.

SUFFICIENCY. Proceeding by reductio ad absurdum, let V be a contraction whose boundary spectrum contains a point λ which is not a root of unity. We can always choose this λ so that $\lambda^m \notin \operatorname{spec} V$ for all $|m| > 1$. We have, given V, to exhibit an orthocomplemented two-dimensional Euclidean subspace. By the Boundary Spectrum Splitting-Off Theorem, we may seek such a subspace in the boundary subspace of the operator V (the relation "M_1 is orthocomplemented in M_2" is transitive. Hence, we may assume that V is an isometry. Consider the closure of the group of operators $\{V^k\}_{k \in \mathbb{Z}}$. It is a compact Abelian group of isometries (the Sushkevich kernel K). Consider its weight χ specified by the condition $\chi(V) = \lambda$. The range of χ contains λ and is a closed subgroup of the unit circle and, as such, it is equal to the latter. Therefore, there is an $A \in K$ such that $\chi(A) = -1$, i.e., $(-1) \in \operatorname{spec} A$. The associated eigensubspace $L \equiv L(A;-1)$ is orthocomplemented (the orthogonal projection onto L is $\lim_{m\to\infty} m^{-1} \sum_{k=0}^{m-1} (-1)^k A^k$) and is invariant under K. Moreover, $A|L = -E$ and $L(V;\lambda) \subset L$. Since $\dim B < \infty$, we may assume that if $C \in K$, $(-1) \in \operatorname{spec}(C|L)$, and $L(V;\lambda) \subset L(C;-1)$, then $C|L = -E$; otherwise, we can reduce L to $L(C;-1)$. Let us show that under these circumstances K does not have in L weights $\theta \neq \chi, \bar{\chi}$. Assume the contrary. Then $\operatorname{spec}(V|L)$ contains a point $\omega \neq \lambda, \bar{\lambda}$. Let Γ denote the closed subgroup of the two-torus $\mathbb{T}^2$ generated by the point (λ,ω). The first coordinate projection $\Gamma \to \mathbb{T}$ is surjective. Hence, Γ contains a point $(-1,\zeta)$, and if the projection is not injective we can take $\zeta \neq -1$. If, however, the projection is injective, we compose its inverse and the second coordinate projection to get a continuous homomorphism $\mathbb{T} \to \mathbb{T}$. But then $\omega = \lambda^m$ for some integer m. For $|m| > 1$ this contradicts the original choice of λ, while for $|m| = 1$ it contradicts the condition $\omega \neq \lambda, \bar{\lambda}$. Finally, for $m = 0$ we get $\omega = 1 \in \mathrm{spec}(V|L)$, whence $1 \in \mathrm{spec}(A|L)$, which is impossible because $A|L = -E$. Thus, $(-1,\zeta) \in \Gamma$ for some $\zeta \neq -1$. Now consider the group homomorphism $U \to (\chi(U),\theta(U))$ of K into Γ; here θ is the weight specified by the condition $\theta(V) = \omega$. It is surjective, and so there is an $U \in K$ such that $\chi(U)$ and $\theta(U)$ are not simultaneously equal to -1. Therefore, $(-1) \in \mathrm{spec}(U|L)$, $L(V;\lambda) \subset L(U;-1)$, but $U|L \neq -E$, which contradicts our assumption.

Thus, L is an orthocomplemented K-invariant subspace, and the weights of K in L are χ and $\bar{\chi}$. L is even-dimensional and possesses a natural complex structure, given by the action of K. This structure is compatible with the norm because the operators in K are isometric.

Now pick a vector $v \in L$, $\|v\| = 1$, and a complex-linear supporting functional f for v : $f(v) = 1$, $\|f\| = 1$. The operator $\Pi = f(\cdot)v$ is the orthogonal projection onto the complex linear span of the vector v, i.e., onto a two-dimensional real K-invariant subspace. The latter is Euclidean since in it multiplication by ζ $(|\zeta| = 1)$ is isometric.

□

<u>Exercise 1.</u> *An equivalent characterization of the spaces in the class described by the theorem is that the group of isometries of any orthocomplemented subspace is finite* (Yu. I. Lyubich, 1970).

<u>Exercise 2.</u> *The class of spaces under consideration contains all finite dimensional real spaces* ℓ_p, $1 \leq p \leq \infty$ (Yu. I. Lyubich, 1970).

Recently A. I. Veitsblit in collaboration with the author have obtained a cone analogue of the preceding theorem. We omit the statement of this result. We remark that the first progress in the set of problems discussed in 6° has been achieved by M. A. Krasnosel'skii (1968).

CHAPTER 5

REPRESENTATIONS OF LOCALLY COMPACT ABELIAN GROUPS

1. ELEMENTS OF HARMONIC ANALYSIS

1°. The harmonic analysis on locally compact Abelian groups is contructed following the model provided by the theory of Fourier integrals (and, needless to say, encompasses the Fourier series and the more general harmonic analysis on compact groups). This domain emerged at the end of the thirties and the beginning of the forties thanks to works of A. Weil, I. M. Gelfand, D. A. Raikov, and M. G. Krein. To this day harmonic analysis has undergone a tremendous development (to get an idea about this the reader should have a look at the two-volume monograph of E. Hewitt and K. Ross [19] ; in the Preface to the 2nd volume the authors write : "Obviously we have not been able to cover all of harmonic analysis"). We restrict ourselves to a modest introduction necessary for the ensuing discussion.

Let G be a locally compact Abelian group, dg a Haar measure on G (normalized, if G is compact), and G^* the dual group, i.e., the group of one-dimensional unitary characters of G endowed with the compact-open topology (unless otherwise stipulated, we will consider here only such characters). If G is compact (discrete), then G^* is discrete (respectively, compact). By the Duality

Principle, G^* is also locally compact, and G^{**} is topologically isomorphic to G ; the canonical isomorphism $i : G \to G^{**}$ is given by $(ig)(\chi) = \chi(g)$. In particular, i is injective, which means precisely that there are sufficiently many characters. In the following we will identify G^{**} and G (the purpose of this identification is to give a clear picture, though we do not use it always, and when we do, by far not in complete manner).

The basic construction of harmonic analysis on a group G is the *Fourier transformation*

$$\tilde{\phi}(\chi) = \phi(g)\overline{\chi(g)}\, dg , \quad \chi \in G^* ,$$

which assigns to each function $\phi \in L_1(G)$ a function $\tilde{\phi}$ on G^*. For fixed χ this defines a multiplicative functional $\phi \to \hat{\chi}(\phi) = \tilde{\phi}(\chi)$ on the group algebra $L_1(G)$. [For fixed χ, $\tilde{\phi}(\chi)$ is the Fourier transform of the function ϕ with respect to the one-dimensional representation χ of G (in the sense of Chap. 3, Sec. 1, 5°), and by the relevant theorem it is an algebra morphism from $L_1(G)$ into $\mathrm{End}\, \mathbb{C} \simeq \mathbb{C}$.] We thus have a *canonical map* of the group G^* into the maximal ideal space $M = M(L_1(G))$ of the group algebra. If G^* is not compact, then G is not discrete, and the unit element j in the group algebra is adjoined formally. Also, $\hat{\chi}$ admits a multiplicative extension by the rule : $\hat{\chi}(\omega j + \phi) = \omega + \hat{\chi}(\phi)$, and the map $\omega j + \phi \to \omega$ is a multiplicative functional, too, the "point at infinity" in M. On the other hand, if G^* is not compact, we can build its one-point compactification $\overline{G}^* = G^* \cup \{\infty\}$. For a compact G^* we put $\overline{G}^* = G^*$. The canonical map $G^* \to M$ extends to $\overline{G}^*$ by the rule $\infty \to \omega$.

THEOREM. *The canonical map* $\hat{}\ : \overline{G}^* \to M$ *is a homeomorphism.*

PROOF. The injectivity of $\hat{}$ is obvious. Its surjectivity follows from the general form of a linear functional on $L_1(G)$ and the multiplicativity requirement, which leads to the functional equation $\psi(gh) = \psi(g)\psi(h)$ (a.e.). Every bounded measurable solution of this equation is continuous (possibly after it is suitably modified on a set of measure zero), i.e., it is a character, and in fact a unitary one, thanks to its boundedness. Since the spaces $\overline{G}^*$ and M are compact, it remains to check that $\hat{}$ is continuous. By the definition of the topology on M, the continuity of $\hat{}|\overline{G}^*$ means the continuity of the Fourier transform $\tilde{\phi}(\chi)$ as a function on G^*, and the latter is readily verified pointwise. In fact, since $\phi \in L_1(G)$, given any $\varepsilon > 0$ one can find a compact $Q \subset G$ such that $|\int_{G \setminus Q} \phi(g)\overline{\chi(g)}\, dg| < \varepsilon$ for all χ. Our task is thus reduced to checking the continuity of the "truncation" $\int_Q \phi(g)\overline{\chi(g)}dg$. The latter follows from the estimate

$$\left|\int_Q \phi(g)\overline{\chi(g)}dg - \int_Q \phi(g)\overline{\chi_0(g)}dg\right| \leqslant \|\phi\| \max_{g\in Q} |\chi(g) - \chi_0(g)|$$

and the definition of the topology on G^*. The continuity of $\hat{}$ at the point ∞ (if it is present) means that $\tilde{\phi}(\chi) \to 0$ as $\chi \to \infty$, i.e., for any $\varepsilon > 0$ there is a compact set $K \subset G^*$ such that $|\tilde{\phi}(\chi)| < \varepsilon$ for all $\chi \notin K$ (for the classical Fourier transformation this is the *Riemann-Lebesgue Theorem*). To show that this is the case, let N denote the neighborhood of the point ω in M defined by the inequality $|\hat{\chi}(\phi) - \omega(\phi)| < \varepsilon$, i.e., $N = \{\hat{\chi} \mid |\tilde{\phi}(\chi)| < \varepsilon\}$. The set $M \setminus N = \{\hat{\chi} \mid |\tilde{\phi}(\chi)| \geqslant \varepsilon\}$ is compact in M. Its preimage in G^* is at any rate closed. Let us show that it is compact, and so it can serve as the set K. It suffices to verify that this set of functions on G is uniformly equicontinuous. To this end we use the self-evident formula

$$\int \{\phi(gh) - \phi(g)\}\overline{\chi(g)}\, dg = \{\chi(h) - 1\}\tilde{\phi}(\chi).$$

It yields the estimate

$$|\chi(h) - 1| \leqslant \varepsilon^{-1}\|R(h)\phi - \phi\|,$$

where $R(h)$ stands, as usual, for translation. Let N be a neighborhood of identity in G such that $\|R(h)\phi - \phi\| < \varepsilon^2$ for all $h \in N$. Then $|\chi(h) - 1| < \varepsilon$, and hence $|\chi(gh) - \chi(g)| < \varepsilon$

for all $g \in G$.

□

From now on we shall (canonically) identify $\overline{G}^*$ with M. The Fourier transformation, viewed as the map $\phi \to \tilde{\phi}$, or $(\lambda j + \omega) \to \lambda + \tilde{\phi}$, depending on whether G is discrete or not, acts from the group algebra into $C(M)$.

COROLLARY. *The Fourier transformation coincides with the Gelfand homomorphism.*

□

Its image is dense in $C(M)$, since the group algebra is symmetric : $\tilde{\overline{\phi}}(\chi) = \phi^*(\chi)$, where $\phi^*(g) = \overline{\phi(g^{-1})}$.

Now let $d\chi$ be a Haar measure on G^*. The Fourier transform $\tilde{\psi}$ of any function $\psi \in L_1(G^*)$ is a function on G:

$$\tilde{\psi}(g) = \int_{G^*} \psi(\chi)\overline{\chi(g)}\, d\chi .$$

Multiplying this equality by $\phi(g)$, with $\phi \in L_1(G)$, and integrating the result with respect to dg, we obtain the following version of the *Parseval equality* :

$$\int_G \phi(g)\tilde{\psi}(g)\, dg = \int_{G^*} \tilde{\phi}(\chi)\psi(\chi)\, d\chi ,$$

from which we can extract further useful information.

THEOREM. *The group algebra* $L_1(G)$ *is semisimple.*

In view of the preceding theorem, this says that if $\tilde{\phi}(\chi) = 0$ for all χ, then $\phi = 0$, i.e., it is the *Uniqueness Theorem* for the Fourier transformation.

PROOF. We have :

$$\int \phi(g)\tilde{\psi}(g)\, dg = 0$$

for all $\psi \in L_1(G^*)$. Moreover, $\int \phi(g)dg = \tilde{\phi}(\mathbb{1}) = 0$. But, as we already know, the functions of the form $\tilde{\psi}$ + const form a dense set in $C(\overline{G})$, and $C(\overline{G})$, in its turn, is dense in $L_\infty(G) = L_1(G)^*$ in the w^*-topology. Thus, every linear functional vanishes on ϕ, and so $\phi = 0$.

□

2°. Now consider the more general group algebra $L(G;\alpha)$ defined by an arbitrary weight α. Recall that a weight is, by definition, a positive, measurable (with respect to the Haar measure), locally bounded function α on G possessing the ring property : $\alpha(g_1g_2) \leqslant \alpha(g_1)\alpha(g_2)$. In what follows we shall assume that $\alpha(g) \geqslant 1$ for all $g \in G$. [Every weight α satisfies the condition $\alpha(e) \geqslant 1$; however, the weight $\alpha(k) = c^k$ $(c > 1)$ on $\mathbb{Z}$, for example, tends to zero as $k \to -\infty$. Notice also that always $\alpha(g)\alpha(g^{-1}) \geqslant 1$.] Then obviously $L(G;\alpha) \subset L_1(G)$ and $\|\phi\|_\alpha \geqslant \|\phi\|_1$, and so the imbedding $i_\alpha : L(G;\alpha) \to L_1(G)$ is a Banach algebra morphism (if G is not discrete both algebras are supplied with a formal unit ; for G discrete the unit of $L_1(G)$ belongs to $L(G;\alpha)$). It induces a canonical continuous map $i_\alpha^* : M \to M_\alpha$ of the associated maximal ideal spaces. i_α^* is injective, since $L(G;\alpha)$ is dense in $L_1(G)$ (to see this, notice that, thanks to the local boundedness of the weight α, every compactly-supported function in $L_1(G)$ belongs to $L(G;\alpha)$). Thus, M is canonically identified with a compact subset of the space M_α. Generally speaking, however, $M \neq M_\alpha$.

<u>Example.</u> Consider the weight $\alpha(k) = a^{|k|}$ (with $a \geqslant 1$) on $\mathbb{Z}$. The maximal ideal space of the algebra $L(\mathbb{Z};\alpha)$ is homeomorphic to the annulus $a^{-1} \leqslant |\zeta| \leqslant a$ in the complex plane $\mathbb{C}$. For $a = 1$ the algebra and its maximal ideal space become $L_1(\mathbb{Z})$ and the unit circle $\mathbb{T}$, respectively.

This example shows that the compactum M_α depends not only on G, but also on the weight α, more precisely, on its growth rate on the exponential scale. The latter is measured by the function $a(g) = \lim\limits_{k\to\infty} \{\alpha(g^k)\}^{1/k}$ (by a theorem of M. Fekete, the limit exists and equals $\inf\limits_k \{\alpha(g^k)\}^{1/k}$). It satisfies the inequalities $1 \leqslant a(g) \leqslant \alpha(g)$. Moreover, $a(g^m) = a(g)^m$ for $m \geqslant 0$,

$a(g_1g_2) \leqslant a(g_1)a(g_2)$, and $a(e) = 1$. We call the function $\ln a(g)$ and the number $\sigma = \sup_g \ln a(g)$ the *growth indicator* and the *exponential type* (*e.t.*), respectively, of the weight α. If α is bounded on G, then $a = \mathbb{1}$ (on a compact group G every weight is bounded ; in any case, $L(G;\alpha) = L_1(G)$ whenever α is bounded).

The multiplicative functionals on $L(G;\alpha)$ different from the "improper" functional ω are canonically identified with (generally speaking, nonunitary) characters χ that satisfy the condition $|\chi(g)| \leqslant C_\chi \alpha(g)$, with $C_\chi = \text{const} > 0$. The latter implies that $|\chi(g^k)| \leqslant C_\chi \alpha(g^k)$, and hence that $|\chi(g)| \leqslant \{C_\chi \alpha(g^k)\}^{1/k}$. Consequently, $|\chi(g)| \leqslant a(g)$. Conversely, this last inequality obviously gives $|\chi(g)| \leqslant \alpha(g)$.

Exercise. *The characters in the class under consideration satisfy the lower bound* $|\chi(g)| \geqslant [a(g^{-1})]^{-1}$.

Remark. *The space* M_α *contains the group* G^*; *and* G^* *acts naturally on* M_α.

In the remaining part of this section we will be concerned only with weights of null e.t., i.e., we will assume that $a = \mathbb{1}$. In this case $M_\alpha \setminus \{\omega\}$ consists of unitary characters. From the foregoing analysis we obtain the following

THEOREM. *Suppose the weight* α *has null e.t., i.e.,* $\lim_k \{\alpha(g^k)\}^{1/k} = 1$ *for all* g. *Then the maximal ideal space* M_α *of the algebra* $L(G;\alpha)$ *is canonically homeomorphic to* $\overline{G}^*$.

□

From now on we shall identify M_α and $\overline{G}^*$.

Concerning the semisimplicity of the Banach algebra $L(G;\alpha)$, it holds without any constraints on the growth of α. In fact, if $\tilde{\phi}(\chi) = 0$ for all $\chi \in M_\alpha$, then a fortiori $\tilde{\phi}(\chi) = 0$ for all $\chi \in M$, and then $\phi = 0$ by the semisimplicity of $L_1(G)$.

3°. In the circle of problems of representation theory that we next address an essential role is played by the *non-quasi-analyticity condition*

$$\sum_{k=1}^{\infty} k^{-2} \ell n\, \alpha(g^k) < \infty \quad \forall\, g \in G.$$

This terminology is connected with the following *Quasi-Analyticity Theorem* (N. Levinson, 1936).

Let $\beta(k) \geqslant 1$ *be a nondecreasing function on* $\mathbb{Z}_+$ *such that* $\sum_{k=1}^{\infty} k^{-2} \ell n\, \beta(k) = \infty$. *Suppose that the Fourier coefficients of the function* $\phi \in L_1(\mathbb{T})$ *satisfy the condition* $\sup_{k\geqslant 1} \beta_k |c_{-k}| < \infty$ *and that* ϕ *vanishes a.e. on an interval. Then* $\phi = 0$.

In the simplest case, where $c_{-k} = 0$ for all $k \geqslant 1$, ϕ admits an analytic continuation to the unit disk and one is reduced to the known uniqueness theorem. The *quasi-analyticity* of one or other class of functions means that a uniqueness theorem holds for this class [not necessarily under *condition I : "ϕ vanishes on an interval"*. The original formulation of J. Hadamard (1932) was concerned with Δ-*quasi-analyticity*, i.e., with the uniqueness of an infinitely differentiable function ϕ such that $\phi^{(n)}(0) = 0$ for $n = 0,1,2,\ldots$. The notion of *I-quasi-analyticity* was introduced by S. N. Bernshtein (1923). Levinson's Theorem generalizes one of the first *I*-quasi-analyticity theorems, due to Ch. de la Vallée-Poussin (1924). The reader interested in the classical results in this domain may consult the book of S. Mandelbrojt [33]].

<u>Exercise.</u> *The e.t. of any non-quasi-analytic weight* α *is equal to zero. Consequently,* $\alpha(g) \geqslant 1$ *for all* $g \in G$.

Our arguments will rely heavily on the following important result.

THEOREM (Y. Domar, 1956). *Let* α *be a non-quasi-analytic weight on the group* G. *Then for every compact set* K G^* *and every neighborhood* $U \supset K$ *there exists a function* $\phi \in L(G;\alpha)$ *whose Fourier transform* $\tilde{\phi}$ *is equal to one on* K *and to zero on outside* U, *and satisfies* $0 \leqslant \tilde{\phi}(\chi) \leqslant 1$ *throughout* G^*.

PROOF. We first solve a less demanding problem : to produce a function $\psi \in L(G;\alpha)$ such that $\tilde{\psi}$ vanishes outside an arbitrarily prescribed neighborhood of the identity $\mathbb{1} \in G^*$ and $\tilde{\psi}(\mathbb{1}) \neq 0$. We let $\mathcal{D}$ denote the class of groups for which this problem can be solved (for any non-quasi-analytic weight α). We claim that $\mathcal{D}$ coincides with the class of all (locally compact Abelian) groups. The proof is broken into several steps which use a number of fundamental facts.

1. If G is compact, then $G \in \mathcal{D}$. In fact, in this case α is bounded, and so $L(G;\alpha) = L_1(G)$ [notice that every bounded weight is non-quasi-analytic]. For $\psi = \mathbb{1} \in L_1(G)$ we have $\tilde{\psi}(\mathbb{1}) = 1$ and $\tilde{\psi}(\chi) = 0$ for all $\chi \neq \mathbb{1}$, thanks to the orthogonality of the unit character on all the others.

2. $\mathbb{R} \in \mathcal{D}$. This, nontrivial in itself fact follows readily from the first part of the well-known *Paley-Wiener Theorem* (see [37], where this theorem serves as the basis for the proof of a criterion for Δ-quasi-analyticity found earlier by T. Carleman (1926).

Let $\mu \in L_2(\mathbb{R})$, $\mu \geqslant 0$, $\mu \neq 0$. If $\int_{-\infty}^{\infty} (1+t^2)^{-1} |\ln \mu(t)| dt < \infty$, then there exists a function $\theta \in L_2(\mathbb{R})$ such that $|\theta(t)| = \mu(t)$ for a.e. t and the Fourier-Plancherel transform $\tilde{\theta}(\lambda)$ vanishes for a.e. $\lambda < 0$. Conversely, if such a function exists, then μ satisfies the indicated condition. [We remind the reader that the Fourier-Plancherel transformation is the natural extension of the Fourier transformation from $L_1(\mathbb{R}) \cap L_2(\mathbb{R})$ to $L_2(\mathbb{R})$. Here we can replace $\mathbb{R}$ by any locally compact Abelian group G. *The Fourier-Plancherel transformation is, up to a constant factor, a Hilbert-space isometry of* $L_2(G)$ *onto* $L_2(G^*)$. In the case $G = \mathbb{R}$

this result is due to M. Plancherel (1910); the general case is due independently to A. Weil (1940) and M. G. Krein (1941). An exposition of the Plancherel Theorem is given, for example, in the book of F. Riesz and B. Sz.-Nagy [41].]

Now notice that if α is a non-quasi-analytic weight on $\mathbb{R}$, then $\int_{-\infty}^{\infty}(1+t^2)^{-1}\ell n\ \alpha(t)\ dt < \infty$. For, if k denotes the integer part of t, then $\alpha(t) \leqslant C\alpha(k)$, with $C = \sup_{0\leqslant s\leqslant 1} \alpha(s)$. The function $\mu(t) = [(1+t^2)\alpha(t)]^{-1}$ satisfies the conditions of the Paley-Wiener Theorem. Consequently, there exists a function $\theta(t)$ with the suitable properties. Since obviously $\theta \in L_1(\mathbb{R})$, the Fourier-Plancherel transform $\tilde{\theta}$ is equal to the ordinary Fourier transform and is continuous. Moreover, $\tilde{\theta}(\lambda) = 0$ for all $\lambda \leqslant 0$. We may assume, with no loss of generality, that $\lambda_0 = \inf\{\lambda \mid \tilde{\theta}(\lambda)\neq 0\} = 0\}$ (otherwise we replace θ by $\theta(t)e^{-i\lambda_0 t}$). Take an arbitrarily small $\varepsilon > 0$ such that $\theta(\varepsilon) \neq 0$. Put $\theta_+(t) = \theta(t)e^{-i\varepsilon t}$, $\theta_-(t) = \overline{\theta(t)}e^{i\varepsilon t}$, and $\psi = \theta_+ * \theta_-$, where $*$ denotes convolution. This is precisely the sought-for function for the ε-neighborhood of zero in $\mathbb{R}^* \equiv \mathbb{R}$. In fact, notice first that

$$\tilde{\psi}(\lambda) = \tilde{\theta}_+(\lambda)\tilde{\theta}_-(\lambda) = \tilde{\theta}(\varepsilon+\lambda)\overline{\tilde{\theta}(\varepsilon-\lambda)} ,$$

and so $\tilde{\psi}(\lambda) = 0$ whenever $|\lambda| \geqslant \varepsilon$. Secondly, $\tilde{\psi}(0) = |\tilde{\theta}(\varepsilon)|^2 \neq 0$. Finally, we show that $\psi \in L(\mathbb{R};\alpha)$. We have $|\psi| \leqslant \mu * \mu$, i.e.,

$$|\psi(t)| \leqslant \int_{-\infty}^{\infty} \frac{ds}{(1+s^2)[1+(t-s)^2]\alpha(s)\alpha(t-s)} \leqslant$$

$$\leqslant \frac{1}{\alpha(t)}\int_{-\infty}^{\infty} \frac{ds}{(1+s^2)[1+(t-s)^2]} = \frac{2\pi}{\alpha(t)(t^2+4)} ,$$

which proves our claim. [V. A. Marchenko (1950) proposed an explicit construction yielding a function ψ which enjoys supplementary useful properties.]

3. $\mathbb{Z} \in \mathcal{D}$. We reduce this case to the preceding one. To this end, we extend α to the full real line, setting

$$\overset{\circ}{\alpha}(t) = \alpha(k) + (t-k)[\alpha(k+1) - \alpha(k)] ,$$

where k designates the integer part of t, and then we put

$$\beta(t) = \sup_s \{\overset{\circ}{\alpha}(t+s)\overset{\circ}{\alpha}(0)/\overset{\circ}{\alpha}(s)\} .$$

The last operation is required in order to restore the ring property, lost through interpolation. Obviously, $\beta(t) \geqslant \overset{\circ}{\alpha}(t) \geqslant 1$. On the other hand, $\beta(t) \leqslant C\alpha(k)$, thanks to the ring property of α (here we can put $C = \max\{\alpha(0),\alpha(1)\}$). Therefore, β is a non-quasi-analytic weight on $\mathbb{R}$.

We construct a function $\psi \in L(\mathbb{R};\beta)$ such that $\tilde{\psi}(\lambda) = 0$ for all $|\lambda| \leqslant \varepsilon$ (where $0 < \varepsilon < \pi$) and $\tilde{\psi}(0) \neq 0$. Using the construction of the preceding case, we can guarantee the bound $|\psi(t)|\beta(t) \leqslant (1+t^2)^{-1}$. Since $|\psi(k)|\alpha(k) \leqslant (1+k^2)^{-1}$ for all $k \in \mathbb{Z}$, the function $\psi|\mathbb{Z} \in L(\mathbb{Z};\alpha)$. Its Fourier transform is the Fourier series

$$\tilde{\psi}_{\mathbb{Z}}(\zeta) = \sum_{k=-\infty}^{\infty}\psi(k)\bar{\zeta}^k \qquad (\zeta \in \mathbb{T}) ;$$

here

$$\frac{1}{2\pi}\int_{-\pi}^{\pi}\tilde{\psi}_{\mathbb{Z}}(e^{i\lambda})e^{ik\lambda}d\lambda = \psi(k) = \frac{1}{2\pi}\int_{-\pi}^{\pi}\tilde{\psi}(\lambda)e^{ik\lambda}d\lambda$$

for all $k \in \mathbb{Z}$. By the Uniqueness Theorem, $\tilde{\psi}_{\mathbb{Z}}(e^{i\lambda}) = \tilde{\psi}(\lambda)$ $(|\lambda| < \pi)$, i.e., $\tilde{\psi}_{\mathbb{Z}}$ vanishes in the complement of the arc $(e^{-i\varepsilon},e^{i\varepsilon})$, and $\tilde{\psi}_{\mathbb{Z}}(1) = \tilde{\psi}(0) \neq 0$.

4. $G_1, G_2 \in \mathcal{D} \Rightarrow G_1 \times G_2 \in \mathcal{D}$.

In fact, let α be a non-quasi-analytic weight on $G_1 \times G_2$. Set $\alpha_1(g_1) = \alpha(g_1,e_2)$ and $\alpha_2(g_2) = \alpha(e_1,g_2)$ for $g_1 \in G_1$ and $g_2 \in G_2$; e_1 and e_2 are the identity elements of G_1 and G_2, respectively. Obviously, α_1 and α_2 are non-quasi-analytic weights on G_1 and G_2. Let $\psi_1 \in L(G_1;\alpha_1)$ and $\psi_2 \in L(G_2;\alpha_2)$ be functions that vanish identically outside neighborhoods N_1

and N_2 of e_1 and e_2, respectively, and such that $\psi_1(e_1) \neq 0$, $\psi_2(e_2) \neq 0$. Put $\psi(g_1 g_2) = \psi_1(g_1)\psi_2(g_2)$. Then ψ vanishes identically in the complement of $N_1 \times N_2$ and $\psi(e_1, e_2) \neq 0$. Moreover, $\psi \in L(G_1 \times G_2; \alpha)$, since $\alpha(g_1, g_2) \leq \alpha_1(g_1)\alpha_2(g_2)$.

5. Let $G = \Gamma \times \mathbb{R}^p \times \mathbb{Z}^q$, where Γ is a compact group, and p,q are arbitrary nonnegative integers. Then $G \in \mathcal{D}$. This is a straightforward consequence of the foregoing analysis.

6. Any (locally compact Abelian) group G belongs to the class $\mathcal{D}$. In fact, let $N \subset G^*$ be an arbitrary precompact neighborhood of the identity. Let $g(\chi)$ be a continuous function such that $g \equiv 0$ in the complement of N and $g(\mathbb{1}) = 1$. Since the image of the algebra $L_1(G)$ under the Gelfand representation is dense in $C(G^*)$, there exists a function $\gamma \in L_1(G)$ with compact support such that $\|g - \tilde{\gamma}\|_{C(G^*)} \leq \frac{1}{3}$. Put $\gamma_0 = \gamma/\tilde{\gamma}(\mathbb{1})$. Then $\tilde{\gamma}_0(\mathbb{1}) = 1$ and $|\tilde{\gamma}_0(\chi)| \leq \frac{1}{2}$ for all $\chi \notin N$. Moreover, there is a symmetric neighborhood of e with compact closure Q such that $\gamma_0(g) = 0$ for all $g \notin Q$. Let $H \subset G$ denote the subgroup generated by Q. By the Structure Theorem (see, for example, Morris's book [34]), $H \approx \Gamma \times \mathbb{R}^p \times \mathbb{Z}^q$ with Γ compact and p,q nonnegative integers. We already know that $H \in \mathcal{D}$. Let M be the neighborhood of identity in H^* specified by the inequality $\left|\int_H \gamma_0(h)\overline{\xi(h)}dh\right| > \frac{1}{2}$, where $dh = dg|H$ is a Haar measure on H. [$\mathbb{1} \in M$, since $\gamma_0(g) = 0$ for for $g \in H$, and so $\int_H \gamma_0(h)dh = \int_G \gamma_0(g)dg = \tilde{\gamma}_0(\mathbb{1}) = 1$.] Let $\psi_0 \in L(H; \alpha|H)$ be a function such that $\tilde{\psi}_0(\xi) = 0$ for all $\xi \notin M$ and $\tilde{\psi}_0(\mathbb{1}) \neq 0$. Let ψ denote the extension of ψ_0 by 0 to the whole G. Obviously, $\psi \in L(G;\alpha)$ and $\tilde{\psi}(\chi) = \tilde{\psi}_0(\chi|H)$. Hence, if $\chi \notin N$, and consequently $\left|\int_G \gamma_0(g)\overline{\chi(g)}dg\right| \leq \frac{1}{2}$, i.e., $\left|\int_H \gamma_0(h)\overline{(\chi|H)(h)}dh\right| \leq \frac{1}{2}$, then $\chi|H \notin M$. This gives $\tilde{\psi}_0(\chi|H) = 0$, i.e., $\tilde{\psi}(\chi) = 0$. Moreover, $\tilde{\psi}(\mathbb{1}) = \tilde{\psi}_0(\mathbb{1}) \neq 0$.

The next step is to infer from the inclusion $G \in \mathcal{D}$ the needed, more precise conclusion. Let $K \subset G^*$ be compact and let $U \supset K$ be open. We have to exhibit a function $\phi \in L(G;\alpha)$ such that $\tilde{\phi}|K = \mathbb{1}$, $\tilde{\phi}|\overline{G}^* \setminus U = 0$, and $0 \leq \tilde{\phi} \leq 1$. Let N be a symmetric precompact neighborhood of the identity in G^* such that $KN^2 \subset U$. Choose a function $\psi \in L(G;\alpha)$ such that $\tilde{\psi}(\chi) = 0$ for $\chi \notin N$ and $\tilde{\psi}(\mathbb{1}) \neq 0$. We may assume, with no loss of generality, that the algebra $L(G;\alpha)$ is symmetric ; indeed, the symmetrized weight $\alpha(g)\alpha(g^{-1}) \geq \alpha(g)$ is non-quasi-analytic together with α. Then the function ψ^*, $\psi^*(g) = \overline{\psi(g^{-1})}$ belongs to $L(G;\alpha)$, and so $\psi * \psi^* \in L(G;\alpha)$; also, $(\psi * \psi^*)^\sim = |\tilde{\psi}|^2$. Therefore, we may assume from the very beginning that $\tilde{\psi} \geq 0$. Also, we can normalize ψ so that $\int_{G^*} \tilde{\psi}(\chi)d\chi = 1$. Let θ denote the indicator function of the set KN. It is readily seen that the convolution $\theta * \tilde{\psi}$ can be expressed as

$$(\theta * \tilde{\psi})(\chi) = \int_G \psi(g)\hat{\theta}(g)\overline{\chi(g)}\,dg ,$$

where

$$\hat{\theta}(g) = \int_{G^*} \theta(\xi)\xi(g)d\xi .$$

The function θ is bounded, and so putting $\phi \equiv \psi\hat{\theta}$ we have $\phi \in L(G;\alpha)$ and $\tilde{\phi} = \theta * \tilde{\psi}$. Since $0 \leq \theta(\chi) \leq 1$ and $\tilde{\psi}(\chi) \geq 0$, it follows that $0 \leq \tilde{\phi}(\chi) \leq 1$. Now let $\chi \in K$. Then

$$\tilde{\phi}(\chi) = \int_{G^*} \theta(\chi\xi^{-1})\tilde{\psi}(\xi)d\xi = \int_N \theta(\chi\xi^{-1})\tilde{\psi}(\xi)d\xi =$$

$$= \int_N \tilde{\psi}(\xi)d\xi = \int_{G^*} \tilde{\psi}(\xi)d\xi = 1 .$$

On the other hand, if $\tilde{\phi}(\chi) \neq 0$, then there is a $\xi \in N$ such that $\chi\xi^{-1} \in KN$, and hence $\chi \in KN^2 \subset U$. Consequently, $\tilde{\phi}(\chi) = 0$ for all $\chi \notin U$.

□

COROLLARY 1. *Let* α *be a non-quasi-analytic weight. Then the Banach algebra* $L(G;\alpha)$ *is regular.*

□

In the case $G = \mathbb{Z}$ this result belongs to G. E. Shilov (1947).

COROLLARY 2. *Let* α *be a non-quasi-analytic weight. Then the set* Φ *of the functions* ϕ $L(G;\alpha)$ *whose Fourier transforms* $\tilde{\phi}$ *have compact support is a dense linear manifold in* $L(G;\alpha)$.

PROOF. Suppose Φ is not dense. Then there exists a function ε on G such that $\sup_g |\varepsilon(g)|/\alpha(g) < \infty$ and $\int_G \phi\varepsilon dg = 0$ for all $\phi \in \Phi$. We already proved that $\Phi \neq 0$. Obviously, Φ is invariant under translations and multiplication by characters. Hence, if $\phi \in \Phi$, $\phi \neq 0$, then

$$\int_G \phi(gh)\varepsilon(g)\overline{\chi(g)}\, dg = 0$$

for all $\chi \in G^*$ and all $h \in G$. Thus, the Fourier transform of the function $\phi(gh)\varepsilon(g)$ vanishes identically. Consequently, for each fixed h, $\phi(gh)\varepsilon(g) = 0$ for a.e. $g \in G$, which implies that $\varepsilon = 0$ a.e.

□

These results will be used in the next section to prove the separability of the spectrum for a certain class of representations.

2. REPRESENTATIONS WITH SEPARABLE SPECTRUM

Let T be a representation of the locally compact Abelian group G. We put, as usual, $\alpha_T(g) = \|T(g^{-1})\|$, and we consider the group algebra $L(G;T) = L(G;\alpha_T)$. On $L(G;T)$ there is defined the Fourier transformation with respect to representation T, $\phi \to \hat{\phi} = \int \phi(g)T(g^{-1})dg$ (here we write $\hat{\phi}$ to avoid confusion with the scalar Fourier transform $\tilde{\phi}$). It is a Banach algebra morphism $L(G;T) \to L(B)$, where B is the representation space. Let $L(T)$ denote the uniform closure of the Fourier image $L_0(T)$ of $L(G;T)$. Since the morphism $L(G;T) \to L(T)$ has a dense image, it induces a homeomorphism of the maximal ideal space $M(L(T))$ onto a compact subset of $\bar{G}^*$. From now on we will identify this compact set with $M(L(T))$ and call it the *L-spectrum of the representation* T. Also, we will refer to $M(L(T)) \setminus \{\omega\}$ as the *finite* L-*spectrum of* T. The Gelfand homomorphism of algebra $L(T)$ sends the operator $\hat{\phi} - \lambda E$ into the function $\chi \to \tilde{\phi}(\chi) - \lambda$ on $M(L(T))$. If $\tilde{\phi}(\chi) - \lambda$ does not vanish, then the operator $\hat{\phi} - \lambda E$ is invertible in $L(T)$, and a fortiori in $L(B)$. Hence, $\lambda \in \operatorname{reg} \hat{\phi}$.

The representation T is called *non-quasi-analytic* if

$$\sum_{k=1}^{\infty} k^{-2} \ln \|T(g^k)\| < \infty \quad \text{for all} \quad g \in G.$$

Obviously, every such T has null exponential type. Therefore, $\operatorname{spec}_a T$ consists of unitary characters.

THEOREM (Yu. I. Lyubich - V. I. Matsaev - G. M. Fel'dman, 1973) *The finite* L-*spectrum of any non-quasi-analytic representation* L *is separable.*

We precede the proof by a number of lemmas interesting in their own right. Unless otherwise stipulated, T is assumed to be non-quasi-analytic.

LEMMA 1. *Let* $\lambda \in \operatorname{reg} \hat{\phi}$. *Then* $\tilde{\phi}(\chi) - \lambda \neq 0$ *for all* $\chi \in M(L(T))$.

PROOF. Let A denote the smallest closed subalgebra of $L(B)$ obtained by adjoining the operator $(\hat{\phi} - \lambda E)^{-1}$ to $L(T)$ (A is obviously commutative). Let $\mathcal{A}$ denote the maximal ideal space of A. Since algebra $L(T)$ is regular (thanks to the regularity of $L(G;T)$), every functional $\chi \in M(L(T))$ extends to a functional $\chi' \in A$. But $\hat{\phi} - \lambda E$ is invertible in A. Consequently, $\chi'(\hat{\phi}) - \lambda \neq 0$, i.e., $\tilde{\phi}(\chi) - \lambda \neq 0$.

□

COROLLARY 1 (Mapping of Spectra Theorem). *The spectrum of the operator* $\hat{\phi}$ *is equal to the range of the restriction of the Fourier transform* $\tilde{\phi}$ *to the* L-*spectrum of representation* T.

□

From this we obtain

COROLLARY 2. *Suppose that* $\tilde{\phi}(\chi) = c = \text{const}$ *in a neighborhood* W *of the* L-*spectrum of* T. *Then* $\hat{\phi} = cE$.

PROOF. Pick a neighborhood $W_1 \supset M(L(T))$ such that $\overline{W}_1 \subset W$. Using again the regularity of $L(G;T)$, find an element $\theta \in L(G;T)$ such that $\tilde{\theta}|M(L(T)) = 1$ and $\tilde{\theta}|\overline{G}^* \setminus W_1 = 0$. Then $(\phi - cj)^{\sim}\tilde{\theta} = 0$, whence $(\phi - cj)^{\wedge}\hat{\theta} = 0$. By Lemma 1, $\hat{\theta}$ is invertible in $L(B)$. Consequently, $(\phi - cj)^{\wedge} = 0$, i.e., $\hat{\phi} = cE$.

□

LEMMA 3. *The representation* T *is uniformly continuous if and only if its* L-*spectrum does not contain the point at infinity* ω, *i.e., if and only if the finite* L-*spectrum of* T *is compact.*

PROOF. NECESSITY. Suppose T is uniformly continuous. Then $T(g) \in L(T)$ for all $g \in G$. Let f be a multiplicative functional on $L(T)$. Then $f(\hat{\phi}) = \int \phi(g) f(T(g^{-1}))dg$. The function $\chi(g) = f(T(g))$ is a unitary character, and $\tilde{\phi}(\chi) = f(\hat{\phi})$. Since $\chi \neq \omega$ and f is arbitrary, it follows that ω does not belong to the L-spectrum of T.

SUFFICIENCY. Suppose $\omega \notin M(L(T))$. Then $M(L(T))$ is compact in G^*. Pick a function $\phi \in L(G;T)$ such that $\tilde{\phi}(\chi) = 1$ in a neighborhood $W \supset M(L(T))$. By Lemma 2, $\hat{\phi} = E$, i.e.,

$\int \phi(g)T(g^{-1})dg = E$. Then $T(h) = \int \phi(gh^{-1})T(g^{-1})dg$ for all h G, and

$$\|T(h) - E\| \leq \int |\phi(gh^{-1}) - \phi(g)|\alpha_T(g)dg \to 0$$

as $h \to e$.

□

PROOF OF THE THEOREM. Let $U \neq \emptyset$ be an open subset of the L-spectrum of T such that $\omega \notin \bar{U}$. We show that the compact set $Q = \bar{U}$ is spectral. Let $\Phi(Q)$ denote the set of all functions $\phi \in L(G;T)$ with the property that $\tilde{\phi}(\chi) = 1$ in some absolute (i.e., refering to $\bar{G}^*$) neighborhood of Q. We claim that $L(Q) = \{x \mid \hat{\phi}x = x \ \forall \phi \in \Phi(Q)\}$ is a spectral subspace and that the L-spectrum σ_Q of the representation $T_Q = T|L(Q)$ is equal to Q. The invariance of $L(Q)$ follows from the fact that each operator $\hat{\phi}$ commutes with $T(g)$ for all $g \in G$. Since $\|T_Q(g)\| \leq \|T(g)\|$, we have $L(G;T) \subset L(G;T_Q)$. Now let $\chi_0 \notin Q$. Pick disjoint neighborhoods $V \supset Q$ and W χ_0, and find a $\psi \in L(G;T)$ such that $\tilde{\psi}(\chi_0) = 1$ and $\tilde{\psi}(\chi) = 0$ for all $\chi \notin W$. Now let $\phi \in \Phi(Q)$ satisfy $\tilde{\phi}(\chi) = 0$ for all $\chi \notin V$. Then $\tilde{\psi}\tilde{\phi} = 0$, whence $\hat{\psi}\hat{\phi} = 0$. Consequently, $\hat{\psi}|L(Q) = 0$. Since $\psi \in L(G;T_Q)$, $\hat{\psi}|L(Q)$ is the Fourier transform of ψ with respect to T_Q. By the Mapping of Spectra Theorem, $\tilde{\psi}(\chi) = 0$ on σ_Q. Furthermore, $\chi_0 \notin \sigma_Q$, because $\tilde{\psi}(\chi_0) = 1$. Hence, $\sigma_Q \subset Q$. Since σ_Q is closed and $\bar{U} = Q$, to prove that $\sigma_Q = Q$ it now suffices to verify that $\sigma_Q \supset U$.

Suppose $\chi_0 \in U$, but $\chi_0 \notin \sigma_Q$. Pick an open set $U_0 \subset \bar{G}^*$ such that $\chi_0 \in U_0$, $\bar{U}_0 \cap \sigma_Q = \emptyset$, and U_0 does not intersect a neighborhood of $M(L(T)) \setminus Q$. Next, choose a function $\psi \in L(G;T)$ such that $\tilde{\psi}(\chi_0) = 1$ and $\tilde{\psi}(\chi) = 0$ in the complement of U_0. Then, as above, $\hat{\psi}|L(Q) = 0$. Now let $\theta \in \Phi(Q)$ satisfy $\tilde{\theta}(\chi) = 1$ in a neighborhood $V \supset Q$. Since $U_0 \cap M(L(T)) \subset Q$, there exists a neighborhood $W \supset M(L(T))$ such that $U_0 \cap W \subset V$. Then $\tilde{\theta}(\chi) = 1$ for all $\chi \in U_0 \cap W$, and so $(1 - \tilde{\theta}(\chi))\tilde{\psi}(\chi) = 0$

for all $\chi \in W$. By Lemma 2, $\hat{\theta}\hat{\psi} = \hat{\psi}$. This shows that $\mathrm{Im}\,\hat{\psi} \subset L(Q)$. Consequently, $\hat{\psi}^2 = 0$. But then $\tilde{\psi}(\chi) = 0$ on $M(L(T))$; in particular, $\tilde{\psi}(\chi_0) = 1$, which contradicts the choice of ψ. This proves that $\sigma_Q = Q$.

It remains to take an arbitrary invariant subspace M such that the L-spectrum of the representation $T|M$ is contained in Q and show that $M \subset L(Q)$. Now, if $\phi \in \Phi(Q)$, then $\tilde{\phi}(\chi) = 1$ in a neighborhood of the compact set σ_Q. By Lemma 2, $\hat{\phi}|M = E$, i.e., $\hat{\psi}x = x$ for all $x \in M$. Hence, $M \subset L(Q)$.

□

From the separability of the finite L-spectrum theorem one can derive separability theorems for other kinds of spectra exploiting the fact that, under certain conditions, they coincide with the finite L-spectrum.

THEOREM (Yu. I. Luybich - V. I. Matsaev - G. M. Fel'dman, 1973). *Suppose the group* G *is separable. Then the* ωa-*spectrum of any non-quasi-analytic representation* T *of* G *is nonempty and separable.*

PROOF. It suffices to show that the ωa-spectrum σ coincides with the finite L-spectrum. Let $\chi_0 \in \sigma$ and let $\{x_k\}$, $\|x_k\| = 1$, be an associated quasi-weight sequence. Then, for any $\phi \in L(G;T)$,

$$\|\hat{\phi}x_k - \tilde{\phi}(\chi_0)x_k\| \leq \int |\phi(g)|\, \|T(g^{-1})x_k - \chi_0(g^{-1})x_k\|\, dg ,$$

and so $\|\hat{\phi}x_k - \tilde{\phi}(\chi_0)x_k\| \to 0$ as $k \to \infty$ by Lebesgue theorem on passing to the limit under the integral sign (this theorem is not valid for nets !). Therefore, the operator $\hat{\phi} - \tilde{\phi}(\chi_0)E$ is not invertible. Consequently, the function $\tilde{\phi}(\chi) - \tilde{\phi}(\chi_0)$ has a zero on $M(L(T))$. By the regularity of the algebra $L(G;T)$ and the arbitrariness of $\phi \in L(G;T)$, this gives that $\chi_0 \in M(L(T))$. Also, $\chi_0 \neq \omega$ from the very beginning.

Conversely, let $\chi_0 \in M(L(T)) \setminus \{\omega\}$, U a precompact neighborhood of χ_0, and $Q = \overline{U}$. To the compact set Q there corresponds a spectral (refering to the L-spectrum) subspace $L(Q)$.

Since the L-spectrum of the representation $T_Q = T|L(Q)$ is equal to Q and so does not contain the point ω, Lemma 3 guarantees that T is uniformly continuous. Next, since G is separable, the ωa-spectrum of T_Q is not empty. Let χ_1 be any point in $\text{spec}_{\omega a}T_Q$. Then, as we saw earlier, χ_1 belongs to the L-spectrum of T_Q. Consequently, $\chi_1 \in Q$. At the same time, $\chi_1 \in \text{spec}_{\omega a}T$. Since $\text{spec}_{\omega a}T$ is closed, it contains χ_0 in view of the arbitrariness of the neighborhood U.

□

The requirement that the group be separable is essential even for ensuring that the ωa-spectrum is not empty. In the example we are familiar with (Chap. 3, Sec. 4, 1°), the representation is trivial, and hence isometric and uniformly continuous. Under no constraints on the group G we have the following

THEOREM. *The a-spectrum of any uniformly continuous non-quasi-analytic representation* T *of* G *is separable.*

PROOF. In fact, in proving the inclusion $M(L(T)) \subset \text{spec}_a T$, we can refer to the fact that the a-spectrum of any uniformly continuous representation is nonempty, and then argue as above. Now let us prove the opposite inclusion. Let $\chi_0 \in \text{spec}_a T$ and $\varepsilon > 0$. Given any $\phi \in L(G;T)$, we choose $K \subset G$ compact such that $\int_{G\setminus K} |\phi(g)|\alpha_T(g)dg < \varepsilon$. Then

$$\|\hat{\phi}x - \tilde{\phi}(\chi_0)x\| \leq \int_K |\phi(g)| \, \|T(g^{-1})x - \chi_0(g^{-1})x\|dg + 2\varepsilon .$$

Since T is uniformly continuous, the family of operators $\{T(g^{-1}) - \chi_0(g^{-1})E\}_{g \in K}$ is compact in the uniform topology. Hence, for any $\eta > 0$ there is a finite collection of points $\{g_i\}_1^m \subset K$ such that $\{T(g_i^{-1}) - \chi_0(g_i^{-1})E\}_1^m$ is an η-mesh in the indicated family. For $\|x\| = 1$ we obtain the estimate

$$\|\hat{\phi}x - \tilde{\phi}(\chi_0)x\| \leq c[\textstyle\sum_{i=1}^m \|T(g_i^{-1})x - \chi_0(g_i^{-1})x\| + m\eta] + 2\varepsilon,$$

where $c = \int_K |\phi(g)|dg$. We take $\eta = \varepsilon/2mc$ and choose the vector x so that

$$\|T(g_i^{-1})x - \chi_0(g_i^{-1})x\| < \eta.$$

Then

$$\|\hat{\phi}x - \tilde{\phi}(\chi_0)x\| < 3\varepsilon,$$

and again we conclude that the operator $\hat{\phi} - \tilde{\phi}(\chi_0)E$ is not invertible.

□

COROLLARY. *The a-spectrum of any non-quasi-analytic representation* T *is nonempty (without any constraints on the group).*

In fact, the restriction of T to any spectral (refering to the L-spectrum) subspace is uniformly continuous.

□

The question of whether the a-spectrum is separable under the only requirement that the representation be non-quasi-analytic is still open.

We next apply the results discussed above to the spectral theory of operators. For every invertible operator $A \in L(B)$ the approximate spectrum coincides with the a-spectrum of the representation $k \to A^k$ of the group $\mathbb{Z}$. The same is valid about spectral subspaces and spectral compact sets. Therefore, we have the following result.

THEOREM (Yu. I. Lyubich - V. I. Matsaev - G. M. Fel'dman, 1973). *Suppose the invertible operator* $A \in L(B)$ *satisfies the condition*

$$\boxed{\sum_{k=-\infty}^{\infty} (k^2 + 1)^{-1} \ln \|A^k\| < \infty .} \qquad (1)$$

Then A *has separable spectrum (and then* spec A *is obviously unimodular).*

□

The existence of nontrivial invariant subspaces under more restrictive (though still close to (1)) conditions was established by J. Wermer (1952).

Using in analogous manner the representation $t \to e^{At}$ of $\mathbb{R}$ we get the following

THEOREM (Yu. I. Lyubich - V. I. Matsaev, 1960). *Suppose the operator* A *satisfies the condition*

$$\int_{-\infty}^{\infty} (t^2+1)^{-1} \ln\|e^{At}\| dt < \infty . \tag{2}$$

Then A *has separable spectrum (and then* spec A *is obviously real).*

□

One can show that conditions (1) and (2) are equivalent to Levinson's condition (see Chap. 1, Sec. 4, 3°) on the resolvent for the unimodular and real spectrum, respectively.

The non-quasi-analyticity condition also guarantees the completeness of the system of spectral subspaces. Moreover, we have the following

THEOREM. *Let* T *be a non-quasi-analytic representation. Then the system of invariant subspaces on which* T *is uniformly continuous is complete.*

PROOF. Let Q be an arbitrary compact subset of the finite L-spectrum of T such that Q is the closure of its interior. With Q there is associated a spectral subspace L(Q) on which T is uniformly continuous. If $\psi \in L(G;T)$ and ψ vanishes identically in the complement of Q, then $\operatorname{Im} \hat{\psi} \subset L(Q)$. In fact, for every function $\phi \in \Phi(Q)$ we have $\tilde{\phi}\tilde{\psi} = \tilde{\psi}$, and so $\tilde{\phi}(\hat{\psi}x) = \hat{\psi}x$ for all x, i.e., $\hat{\psi}x \in L(Q)$. Hence, it suffices to prove the completeness of the system of subspaces $\operatorname{Im} \hat{\psi}$ parametrized by all possible pairs Q, ψ .

Let $f \in B^*$ and $f(\hat{\psi}x) = 0$, i.e.,

$$\int \psi(g) f(T(g^{-1})x) dg = 0.$$

Since ψ is arbitrary, we obtain $f(T(g^{-1})x) = 0$ a.e. . By continuity, $f(T(g^{-1})x) = 0$ everywhere. In particular, $f(x) = 0$, i.e., $f = 0$.

□

In the Lyubich-Matsaev-Fel'dman work repeatedly referred to above, as well as in subsequent publications concerned with separability of the spectrum (G. M. Fel'dman, 1972 ; V. I. Lomonosov, 1979) one can find sharper statements on the completeness of the family of spectral subspaces of a non-quasi-analytic representation. We omit the formulation of these statements.

To conclude we remark that the non-quasi-analyticity condition is in a certain sense necessary for the separability of the spectrum of a representation. Specifically, we have

THEOREM (Yu. I. Lyubich - V. I. Matsaev - G. M. Fel'dman, 1973). *Suppose the weight* α *on the group* G *has null exponential type and there is a* $g_0 \in G$ *such that*

$$\sum_{k=-\infty}^{\infty} (k^2 + 1)^{-1} \ln \alpha(g_0^k) = \infty . \tag{3}$$

Then the L-spectrum of the regular representation R *of* G *in the space* $L(G;\alpha)$ *is not separable. At the same time,* $\|R(g)\| \leq \alpha(g)$ *for all* $g \in G$.

PROOF. In the present situation the L-spectrum of the representation coincides with the dual group G^*. Moreover, condition (3) implies the existence of a compact set $Q \subset G^*$ with nonempty interior such that if $\phi \in L(G;\alpha)$ and $\tilde{\phi}$ vanishes identically in the complement of Q, then $\phi = 0$ (Y. Domar, 1956).

Now suppose that the spectrum of R is separable. Consider the subspace $L(Q)$. The restriction $R(Q)$ of R to $L(Q)$ is uniformly continuous. Its spectrum $\sigma(R(Q))$ is contained in Q and contains the interior points of Q, i.e., $\sigma(R(Q)) \neq \emptyset$ (as a consequence, $L(Q) \neq 0$). Let $\chi_0 \in G^* \setminus Q$. Then $\chi_0 \notin \sigma(R(Q))$. Therefore (see the theorem in Chap. 3, Sec. 4, 2°), there exists an operator of the form

$$A = \sum_{j=1}^{m} \overline{\chi_0(g_j)} \{R(g_j) - \chi_0(g_j)E\}$$

such that A is invertible in the algebra Lin T. Pick an arbitrary function $\phi \in L(G;\alpha)$, $\alpha \neq 0$, and set $\phi_1 = A^{-1}\phi$. Then

$$\sum_{j=1}^{m} \overline{\chi_0(g_j)}\{\phi_1(gg_j^{-1}) - \chi_0(g_j)\phi_1(g)\} = \phi(g) .$$

Fourier-transforming this equality, we get $\tilde{\phi}(\chi_0) = 0$. Thus, $\tilde{\phi}$ vanishes identically in the complement of Q, and so $\phi = 0$, contrary to the choice of ϕ.

□

REFERENCES

Monographs and Textbooks

[1] N. I. Akhiezer and I. M. Glazman: *Theory of Linear Operators in Hilbert Space*, Vol. 1, "Vyshcha Shkola", Khar'kov, 1977; English transl.: Ungar, New York, 1961.

[2] S. Banach: *Théorie des Opérations Linéaires*, Hafner Pub. Co., New York, 1932.

[3] H. A. Bohr: *Almost Periodic Functions*, Chelsea Pub. Co., New York, 1947.

[4] N. Bourbaki: *Éléments de Mathématique. Livre* VI. *Integration*. Chap. 6, *Integration Vectorielle*, Hermann, Paris, 1959; Chap. 7, *Mesure de Haar*, Chap. 8, *Convolutions et Répresentations*, Hermann, Paris, 1963.

[5] N. Bourbaki: *Éléments de Mathématique. Théories Spectrales*, Hermann, Paris, 1967.

[6] T. Carleman: *Les Fonctions Quasi-Analytiques*, Collection Borel, Gauthier-Villars, Paris, 1926.

[7] A. H. Clifford and G. B. Preston: *The Algebraic Theory of Semigroups*, Vol. 1, Math. Surveys No. 7, Amer. Math. Soc., Providence, R.I., 1961.

[8] I. Colojoară and C. Foiaş: *Theory of Generalized Spectral Operators*, Gordon and Breach, New York, London, Paris, 1968.

[9] I. P. Cornfeld, Ya. G. Sinai, and S. V. Fomin: *Ergodic Theory*, "Nauka", Moscow, 1980; English transl.: Springer Verlag, Berlin, Heidelberg, New York, 1982.

[10] R. Courant and D. Hilbert: *Methoden der Mathematischen Physik*, Vol. 1, Springer-Verlag, Berlin, 1931; English transl.: Interscience Publishers, New York, 1953.

[11] M. M. Day: *Normed Linear Spaces*, Springer-Verlag, Berlin, Götingen, Heidelberg, 1958.

[12] J. Diximier: *Les C*-Algèbres et leurs Représentations*, Gauthier-Villars, Paris, 1964; English transl.: North Holland Math. Library, Vol. 15, Amsterdam, 1977.

[13] N. Dunford and J. T. Schwartz: *Linear Operators. Part III, Spectral Operators*, Pure Appl. Math., Vol. 7, Wiley-Interscience, New York, 1971.

[14] F. R. Gantmakher: *Theory of Matrices*, "Nauka", Moscow, 1967; English transl. of 1st ed.: Chelsea Pub. Co., New York, 1959.

[15] I. M. Gelfand, D. A. Raikov, and G. E. Shilov: *Commutative Normed Rings*, "Fizmatgiz", Moscow, 1960; English transl.: Chelsea Pub. Co., New York, 1964.

[16] F. Greenleaf: *Invariant Means on Topological Groups*, Math. Studies, Vol. 16, Van Nostrand Reinhold Co., New York, 1969.

[17] P. Halmos: *Measure Theory*, Van Nostrand, New York, 1950.

[18] P. Halmos: *Lectures on Ergodic Theory*, Chelsea Pub. Co., New York, 1959.

[19] E. Hewitt and K. A. Ross: *Abstract Harmonic Analysis*, Vol. I, Springer-Verlag, Berlin, Heidelberg, New York, 1963; Vol. II, Springer-Verlag, Berlin, Heidleberg, New York, 1970.

[20] E. Hille and R. S. Phillips: *Functional Analysis and Semigroups*, Amer. Math. Soc. Colloq. Publ., Vol. 31, rev. ed., Amer. Math. Soc., Providence, R.I., 1957.

[21] I. Kaplansky, *An Introduction to Differential Algebra*, Hermann, Paris, 1957.

[22] J. L. Kelley: *General Topology*, Van Nostrand, New York, 1955.

[23] A. A. Kirillov: *Elements of the Theory of Representations*, "Nauka", Moscow, 1972; English transl.: Springer-Verlag, Heidelberg, Berlin, New York, 1976.

[24] A. N. Kolmogorov and S. V. Fomin: *Elements of the Theory of Functions and Functional Analysis*, rev. ed., "Nauka", Moscow, 1972; English transl. of prevoius ed.: Prentice-Hall, Englewood Cliffs, N. J., 1970.

[25] A. I. Kostrikin: *Introduction to Algebra*, "Nauka", Moscow, 1980. (Russian).

[26] A. I. Kostrikin and Yu. I. Manin: *Linear Algebra and Geometry*, Izd.-vo Mosk. Gosud. Univ.-ta, Moscow, 1980. (Russian).

[27] M. G. Krein: *Lectures on the Theory of Stability of Solutions of Differential Equations in Banach Space*, Inst. Mat. Akad. Nauk Ukr. SSR, Kiev, 1964. (Russian).

[28] A. G. Kurosh: *Lectures on General Algebra*, "Fizmatgiz", Moscow, 1962. (Russian).

[29] S. Lang: *Algebra*, Addison-Wesley, Reading, Mass., 1965.

[30] N. Levinson: *Gap and Density Theorems*, Amer. Math. Soc. Colloq. Publ., Vol. 26, Amer. Math. Soc., Providence, R.I., 1940.

[31] B. M. Levitan: *Almost Periodic Functions*, "Gostekhizdat", Moscow, 1953. (Russian).

[32] L. H. Loomis: *An Introduction to Abstract Harmonic Analysis*, Van Nostrand, Princeton, N.J., 1953.

[33] S. Mandelbrojt: *Séries de Fourier at Classes Quasi-Analytiques de Fonctions*, Gauthier-Villars, Paris, 1935.

[34] S. Morris: *Pontryagin Duality and the Structure of Locally Compact Abelian Groups*, Cambridge Univ. Press, Cambridge, 1977.

[35] M. A. Naimark: *Normed Rings*, "Nauka", Moscow, 1968; English transl. of 1st ed.: P. Noordhoff, Groningen, 1964.

[36] J. von Neumann: *Mathematical Foundations of Quantum Mechanics*, transl. from the German, Princeton Univ. Press, Princeton, N.J., 1955.

[37] R. E. A. C. Paley and N. Wiener, *Fourier Transforms in the Complex Domain*, Amer. Math. Soc., New York

[38] L. S. Pontryagin: *Continuous Groups*, GONTI, Moscow, Leningrad, 1938; English transl. of 2nd ed.: *Topological Groups*, Gordon and Breach, New York, 1966.

[39] M. M. Postnikov: *Lie Groups and Algebras*, "Nauka", Moscow, 1982. (Russian).

[40] D. A. Raikov: *Harmonic Analysis on Commutative Groups with Haar Measure and the Theory of Characters*, Trudy Mat. Inst. Steklov 14, Izd.-vo Akad. Nauk SSSR, Moscow, Leningrad, 1945. (Russian).

[41] F. Riesz and B. Sz.-Nagy: *Lecons d'Analyse Fonctionnelle*, 6th ed., Akadémiai Kiadó, Budapest, 1972.

[42] S. Rolewicz, *Metric Linear Spaces*, PWN, Warsaw, 1972.

[43] W. Rudin: *Principles of Mathematical Analysis*, McGraw-Hill, New York, 1964.

[44] W. Rudin: *Functional Analysis*, McGraw-Hill, New York, 1973.

[45] J.-P. Serre: *Représentations Linéaires des Groupes Finis*, 2nd ed., Hermann, Paris, 1971; English transl.: Springer-Verlag, Berlin, Heidelberg, New York, 1977.

[46] B. V. Shabat: *Introduction to Complex Analysis*, Vol. I, "Nauka", Moscow, 1976. (Russian).

[47] G. E. Shilov: *On Regular Normed Rings*, Trudy Mat. Inst. Steklov 21, Izd.-vo Akad. Nauk SSSR, Moscow, Leningrad, 1947. (Russian).

[48] S. Ulam: *A Collection of Mathematical Problems*, Interscience Publ. Inc., New York, 1960.

[49] A. Weil: *L'intégration dans les Groupes Topologiques et ses Applications*, Hermann, Paris, 1940.

[50] D. P. Zhelobenko: *Compact Lie Groups and their Representations*, "Nauka", Moscow, 1970; English transl.: Transl. Math. Monographs Vol. 40, Amer. Math. Soc., Providence, R.I., 1973.

Journal Articles

Amerio, L.: *Abstract almost periodic functions and functional equations*, Boll. Unione Mat. Ital. 20 (1965), 267-333.

Aronszjan, N. and Smith, K. T.: *Invariant subspaces of completely continuous operators*, Ann. Math. (2) 60 (1954), 345-350.

Auerbach, G.: *Sur les groupes lineaires*, Studia Math. 6 (1935), 113-127 and 158-166.

Bernstein, S. N.: *Sur les fonctions quasi-analytiques*, C.R. Acad. Sci. Paris 177 (1923), 937-940.

Bernstein, A. R. and Robinson, A.: *Solution of an invariant subspace problem of K. T. Smith and P. R. Halmos*, Pacific J. Math. 16 (1966), 421-432.

Birkhoff, G. A.: *A note on topological groups*, Compositio Math. 3 (1936), 427-430.

Bishop, E.: *A duality theorem for an arbitrary operator*, Pacific J. Math. 9 (1959), 379-397.

Bochner, S.: *Beiträge zur Theorie der fastperiodischen Funktionen, I*, Math. Ann. 96 (1927), 119-147.

Bochner, S.: *Abstrakte fastperiodische Funktionen*, Acta Math. 61 (1933), 149-184.

Bogolyubov, N. N. and Krein, S. G.: *On positive completely continuous operators*, Zh. Inst. Mat. Akad. Nauk Ukr. SSR 9 (1947), 130-139. (Ukrainian).

Bohl, P.: *Ueber die Darstellung von Funktionen einer Variablen durch trigonometrische Reihen mit mehrerer einer Variablen proportionalen Argumenten*, Magister dissertation, Dorpat (1893).

Bohl, P.: *Ueber eine Differentialgleichung der Störungstheorie*, Crelles J. 131 (1906), 268-321.

Bohl, P.: *Ueber ein in der Theorie der säkularen Störungen vorkommends Problem*, J. reine und angew. Math. 135 (1909), 189-283.

Bohr, P.: *Ueber fastperiodische ebene Bewegungen*, Comm. Math. Helv. 4 (1932), 51-46.

Boles Basit, R.: *A generalization of two theorems of M. I. Kadets on indefinite integrals of almost periodic functions*, Mat. Zametki 9 (1971), 311-321; English transl.: Mat. Notices 9 (1971), 181-186.

Branges, L. de: *The Stone-Weierstrass theorem*, Proc. Amer. Math. Soc. 10 (1959), 822-824.

Clifford, A. H.: *Semigroups containing minimal ideals*, Amer. J. Math. 70 (1948), 521-526.

Dixmier, J.: *Les moyennes invariantes dans les semigroupes et leurs applications*, Acta Sci. Math. Szeged 12 (1950), 213-227.

Domar, Y.: *Harmonic analysis based on certain commutative Banach algebras*, Acta Math. 96 (1956), 1-66.

Domar, Y. and Lindahl, L.-A.: *Three spectral notions for representations of commutative Banach algebras*, Ann. Inst. Fourier, Grenoble 25 (1975),1-32.

Doss, R.: *On bounded functions with almost periodic differences*, Proc. Amer. Math. Soc. 12 (1961), 488-489.

Esclangon, E.: *Les fonctions quasi-periodiques*, These, Paris (1904).

Fel'dman, G. M.: *On the basis property of the system of eigensubspaces of an isometric representation*, Fiz. Tekhn. Inst. Nizk. Temp. Akad. Nauk Ukr. SSR, Mat. Fiz. i Funkts. Analiz 3 (1972), 77-80. (Russian).

Fréchet, M.: *Les fonctions asymptotiquement presque-periodiques*, Rev. Sci. 79 (1941), 341-354.

Frobenius, G.: *Über endliche Gruppen*, Sitzungsber. Preuss Akad. Wiss. Berlin (1895), 163-194.

Gelfand, I. M.: *On normed rings*, Dokl. Akad. Nauk SSSR 23 (1939), 430-432. (Russian).

Gelfand, I. M. and Raikov, D. A.: *On the theory of characters of commutative topological groups*, Dokl. Akad. Nauk SSSR 28 (1940), 195-198. (Russian).

Glazman, E. I.: *On characters of compact semigroups*, Teoriya Funktsii, Funkts. Analiz i ikh Prilozehn. 21 (1974) 122-124. (Russian).

Gleason, A.: *Groups without small subgroups*, Ann. Math. 56 (1952) 193-212.

Gorin, E. A.: *A function-algebra variant of a theorem of Bohr-van Kampen*, Mat. Sb. 82, No. 2 (1970), 260-272; English transl.: Math. USSR Sbornik 11, No. 2 (1970), 233-243.

Gorin, E. A. and Lin, V. Ya.: *The topological meaning of Bohr's theorem on the argument of an almost periodic function*, Proc. Fifth All-Union Topological Conference, Novosibirsk (1967). (Russian).

Gurevich, A.: *Unitary representations in Hilbert space of a compact topological group*, Mat. Sb. 13 (1943), 79-86. (Russian).

Haar, A.: *Der Massbegriff in der Theorie der kontinuierlichen Gruppen*, Ann. Math. 34 (1933), 147-169.

Hadamard, J.: *Sur la generalisation de la notion de fonction analytique*, C. R. Seances Soc. Math. France 40 (1912), p. 28.

Halmos, P. R.: *Invariant subspaces of polynomially compact operators*, Pacific J. Math. 16 (1966), 433-438.

Hewitt, E. and Zuckerman, H. S.: *The ℓ_1-algebra of a commutative semigroup*, Trans. Amer. Math. Soc. 83 (1956), 70-97.

Hurwitz, A.: *Über die Erzeugung der Invarianten durch Integration*, Gött. Nachr. (1897), 71-90.

Jacobs, K.: *Ergodentheorie und fastperiodische Funktionen auf Halbgruppen*, Math. Z. 64 (1956), 289-338.

Jacobs, K.: *Fastperiodizitätseigenschaften allgemeiner Halbgruppen in Banach räumen*, Math. Z. 67 (1957), 83-92.

Jamison, B.: *Asymptotic behavior of successive iterates of continuous functions under a Markov operator*, J. Math. Anal. Appl. 9 (1964), 203-214.

Jamison, B. and Sine, R.: *Irreducible almost periodic Markov operators*, J. Math. Mech. 18 (1969), 1043-1057.

Jessen, B. and Tornehave, H.: *Mean motion and zeros of almost periodic functions*, Acta Math. 77 (1945), 137-279.

Kadets, M. I.: *On the integration of almost periodic functions with values in a Banach space*, Funktsion. Anal. Prilozhen. 3, No. 3 (1969), 71-74; English transl.: Function. Anal. Appl. 3, No. 3 (1969), 228-230.

Kakutani, S.: *Über die Metrization der topologishen Gruppen*, Proc. Imp. Acad. Tokyo 12 (1936), 82-84.

Kalyuzhnyi, V. N.: *Commutative groups of isometries of Minkowski spaces*, Sib. Mat. Zh. 15, No. 5 (1974), 1138-1142; English transl.: Sib. Mat. J. 15, No. 5 (1974), 801-803.

Kalyuzhnyi, V. N.: *Quasi-finite groups of isometries of Minkowski spaces*, Teor. Funkts. Funktsion. Anal. i ikh Prilozhen. 29 (1978), 41-49. (Russian).

Kampen, E. R. van: *Locally compact Abelian groups and their character groups*, Ann. Math. 36 (1935), 448-463.

Kampen, E. R. van: *On almost periodic functions of constant absolute value*, J. London Math. Soc. 12 (1937), 3-6.

Krasnosel'skii, M. A.: *On a certain spectral property of linear completely continuous operators in a space of continuous functions*, Probl. Mat. Anal. Slozhnykh Sistem 2 (1968), 68-71.

Krein, M. G.: *Sur une généralisation du théorème de Plancherel au cas des intégrales Fourier sur les groupes topologiques commutatifs*, Dokl. Akad. Nauk SSSR 30 (1941), 484-488.

Krein, M. G.: *A principle of duality for a bicompact group and a square block-algebra*, Dokl. Akad. Nauk SSSR 69 (1949), 726-728. (Russian).

Krein, M. G. and Rutman, M. A.: *Linear operators leaving invariant a cone in a Banach space*, Usp. Mat. Nauk 3, No. 1 (1948), 3-95; English transl.: Amer. Math. Soc. Transl. (1) 10 (1962), 199-325.

Leeuw, K. de and Glicksberg, I.: *Applications of almost periodic compactifications*, Acta Math. 105 (1961), 63-97.

Leeuw, K. de and Glicksberg, I.: *The decomposition of certain group representations*, J. Analyse Math. 15 (1965), 135-192.

Levinson, N.: *On a class of non-vanishing functions*, Proc. London Math. Soc. 41 (1936), 393-396.

Lomonosov, V. I.: *Invariant subspaces for the family of operators which commute with a completely continuous operator*, Funktsion. Anal. Prilozhen. 7, No. 3 (1973), 55-56; English transl.: Function. Anal. Appl. 7, No. 3 (1973), 213-214.

Lomonosov, V. I.: *Some problems of the theory of invariant subspaces*, Author's abstract of candidate in Phys.-Math. Sciences dissertation, Khar'kov (1973). (Russian).

Lomonosov, V. I.: *The joint approximate spectrum of a commutative family of operators*, Teor. Funkts. Funktsion. Anal. i ikh Prilozhen. 32 (1979), 39-47. (Russian).

Lomonosov, V. I., Lyubich, Yu. I., and Matsaev, V. I.: *Duality of spectral subspaces and conditions for the separation of the spectrum of a bounded linear operator*, Dokl. Akad. Nauk SSSR 216 (1974), 737-739; English transl.: Soviet Math. Dokl. 15, No. 1 (1974), 878-881.

Lorch, E. R.: *Means of iterated transformations in reflexive Banach spaces*, Bull. Amer. Math. Soc. 45 (1939), 217-234.

Lyubarskii, G. Ya.: *On the mean integration of almost periodic functions on a topological group*, Usp. Mat. Nauk 3, No. 3 (1948), 195-201. (Russian).

Lyubich, M. Yu.: *The maximum-entropy measure of a rational endomorphism of the Riemann sphere*, Funktsion. Anal. Prilozhen. 16, No. 4 (1982), 78-79; English transl.: Function. Anal. Appl. 16, No. 4 (1982), 309-311.

Lyubich, Yu. I.: *Almost periodic functions in the spectral analysis of an operator*, Dokl. Akad. Nauk SSSR 132, No. 3 (1960), 518-520 (1960); English transl.: Soviet Math. Dokl. 1, No. 3 (1960), 593-595.

Lyubich, Yu. I.: *Conditions for the completeness of the system of eigenvectors of a correct operator*, Usp. Mat. Nauk 18, No. 1 (1963), 165-171. (Russian).

Lyubich, Yu. I.: *On the boundary spectrum of contractions in Minkowski spaces*, Sib. Mat. Zh. 11, No. 2 (1970), 358-369; English transl.: Sib. Math. J. 11, No. 2 (1970), 271-279.

Lyubich, Yu. I.: *On the spectrum of a representation of an Abelian group*, Dokl. Akad. Nauk SSSR 200, No. 4 (1971), 777-780; English transl.: Soviet Math. Dokl. 12, No. 5 (1971), 1482-1486.

Lyubich, Yu. I. and Matsaev, V. I.: *On the spectral theory of linear operators in Banach spaces*, Dokl. Akad. Nauk SSSR 131, No. 1 (1960), 21-23; English transl.: Soviet Math. Dokl. 1, No. 2 (1960), 184-186.

Lyubich, Yu. I. and Matsaev, V. I.: *On operators with separable spectrum*, Mat. Sb. 56, No. 4 (1962), 433-468; English transl.: Amer. Math. Soc. Transl. (2) 47 (1965), 89-129.

Lyubich, Yu. I., Matsaev, V. I., and Fel'dman, G. M.: *On the separability of the spectrum of a representation of a locally compact Abelian group*, Dokl. Akad. Nauk SSSR 201, No. 6 (1971), 1282-1284; English transl.: Soviet Math. Dokl. 12, No. 6

(1971), 1824-1827.

Lyubich, Yu. I., Matsaev, V. I., and Fel'dman, G. M.: *On representations with separable spectrum*, Funktsion. Anal. Prilozhen. 7, No. 2 (1973), 52-61; English transl.: Function. Anal. Appl. 7, No. 2 (1973), 129-136.

Lyubich, Yu. I. and Tabachnikov, M. I.: *Subharmonic functions on a directed graph*, Sib. Mat. Zh. 10, No. 3 (1969), 600-613; English transl.: Sib. Math. J. 10, No. 3 (1969), 432-442.

Maak, W.: *Eine neue Definition der fastperiodischen Funktionen*, Abh. Math. Sem. Hamburg Univ. 11 (1938), p. 240.

Maak, W.: *Periodizitätseigenchaften unitären Gruppen*, Math. Scand. 2 (1954), 334-344.

Marchenko, V. A.: *On some questions concerning the approximation of continuous functions on the whole real line, III*. Zap. Mat. Otd. Fiz.-Mat. Fak.-ta Khar'kov. Univ.-ta i Khar'kov. Mat. Ob.-va 22 (1950), 115-125. (Russian).

Markov, A. A.: *On free topological groups*, Dokl. Akad. Nauk SSSR 31 (1941), 299-302. (Russian).

Mazur, S.: *Sur les anneaux lineaires*, C. R. Acad. Sci. Paris 207 (1938), 1025-1027.

Montgomery, D. and Zippin, L.: *Small subgroups of finite-dimensional groups*, Ann. Math. 56 (1952), 213-241.

Nessonov, N. I.: *Description of representations of a group of invertible operators of a Hilbert space that contain the identity representation of the unitary group*, Funktsion. Anal. Prilozhen. 17, No. 1 (1983), 79-80; English transl.: Function. Anal. Appl. 17, No. 1 (1983), 64-66.

Neumann, J. von: *Zur allgemeinen Theorie des Masses*, Fund. Math. 13 (1929), 73-116.

Neumann, J. von: *Proof of the quasi-ergodic hypothesis*, Proc. Nat. Acad. Sci. USA 18 (1932), 70-82.

Neumann, J. von: *Die Einführung analytischer Parameterin toplogischen Gruppen*, Ann. Math. 34 (1933), 170-190.

Neumann, J. von: *Almost periodic functions in a group, I*. Trans. Amer. Math. Soc. 36 (1934), 445-492.

Neumann, J. von: *Zum Haarschen Mass in topologischen Gruppen*, Compos. Math. 1 (1934), 106-114.

Neumann, J. von: *The uniqueness of Haar's measure*, Mat. Sb. 1 (1936), 721-734.

Numakura, K.: *On bicompact semigroups*, Math. J. Okayama Univ. 1 (1952), 99-108.

Ol'shanskii, A. Yu.: *On the problem of the existence of an invariant mean on a group*, Usp. Mat. Nauk 35, No. 4 (1980), 199-200; English transl.: Russian Math. Surveys 35, No. 4 (1980), 180-181.

Perron, O.: *Jacobischer Kettenbruchalgorithmus*, Math. Ann. 64 (1907), 1-76.

Peter, F. and Weyl, H.: *Die Vollständigkeit der primitiven Darstellungen einer geschlossenen kontinuierlichen Gruppe*, Math. Ann. 97 (1927), 737-755.

Plancherel, M.: *Contribution a l'etude de la representation d'une fonction arbitraire par les integrales definies*, Rend. Circ. Mat. Palermo 30 (1910), 289-335.

Pontrjagin, L. S.: *Les fonctions presque periodiques et l'analisis situs*, C. R. Acad. Sci. Paris 196 (1933), 1201-1203.

Pontrjagin, L. S.: *The theory of topological commutative groups*, Ann. Math. 35 (1934), 361-388.

Pontrjagin, L. S.: *Sur les groupes abeliens continus*, C. R. Acad. Sci. Paris 148 (1934), 328-330.

Pontrjagin, L. S.: *Sur les groupes topologiques compacts et le cinquieme probleme de M. Hilbert*, C. R. Acad. Sci. Paris 198 (1934), 238-240.

Rees, D.: *On semi-groups*, Proc. Camb. Phil. Soc. 38 (1940), 387-400.

Rosen, W. G.: *On invariant means over compact semigroups*, Proc. Amer. Math. Soc. 7 (1957), 1076-1082.

Rosenblatt, M.: *Equicontinuous Markov operators*, Teor. Ver. i Prim. 9 (1964), 205-222.

Schreier, O.: *Abstrakte continuierliche Gruppen*, Abh. Math. Sem. Univ. Hamb. 4 (1925), 15-32.

Shilov, G. E.: *On the extension of maximal ideals*, Dokl. Akad. Nauk SSSR 29 (1940), 83-85. (Russian).

Slodkowski, Z. and Zelasko, W.: *On joint spectra of commuting families of operators*, Studia Math. 50 (1974), 127-148.

Stone, M.: *Applications of the theory of Boolean rings to general topology*, Trans. Amer. Math. Soc. 41 (1937), 375-481.

Suschkewitsch (Sushkevich), A. K.: *Über die endlichen Gruppen onhe das Gezetz der eindeutigen Umkehrbarkeit*, Math. Ann. 99 (1928), 30-50.

Sz.-Nagy, B.: *On uniformly bounded linear transformations in Hilbert space*, Acta Sci. Math. 3 (1947), 152-157.

Taussky, O.: *A reccuring theorem on determinants*, Amer. Math. Monthly 56 (1949), 672-676.

Valée Poussin, de la C.: *Quatre lecons sur les fonctions quasi-analytiques d'une variable réelle*, Bull. Soc. Math. France, 52 (1924), 175-195.

Wermer, J.: *The existence of invariant subspaces*, Duke Math. J. 19 (1952), 615-622.

Weyl, H.: *Theorie der Darstellung kontinierlichen halbeinfacher Gruppen durch lineare Transformationen*, Math. Z. 23 (1925), 271-309.

Weyl, H.: *Integralgleichungen und fastperiodische Funktionen*, Math. Ann. 97 (1927), 338-356.

Weyl, H.: *Mean Motion*, Amer. J. Math. 60 (1939), 889-896 and 61 (1939), 143-144.

Yosida, K. and Kakutani, S.: *Operator-theoretical treatment of Markoff processes and mean ergodic theorem*, Ann. Math. 42 (1941), 188-228.

INDEX

Integral Equations and Operator Theory

The journal is devoted to the publication of current research in integral equations, operator theory and related topics, with emphasis on the linear aspects of the theory. The very active and critical editorial board takes a broad view of the subject and puts a particularly strong emphasis on applications. The journal contains two sections, the main body consisting of refereed papers, and the second part containing short announcements of important results, open problems, information, etc. Manuscripts are reproduced directly by a photographic process, permitting rapid publication.

Subscription Information
1988 subscription
Volume 11 (6 issues)
ISSN 0378-620X

Published bimonthly
Language: English